Agricultural Economics - Current Issues

Edited by Surendra N. Kulshreshtha

Published in London, United Kingdom

IntechOpen

Supporting open minds since 2005

Agricultural Economics - Current Issues
http://dx.doi.org/10.5772/intechopen.78437
Edited by Surendra N. Kulshreshtha

Contributors
Golam Rabbani, Samia Saif, Sirazoom Munira, Boubaker Dhehibi, Mohamed Ben Salah, Aymen Frija, Kala Brigitte Hema, Bienlo Annick Marina Paré, Marie-Thérèse Arcens-Somé, Wolfgang Johann von Loeper, Scott Drimie, James Blignaut, Ali Chebil, Mariem Makhlouf, Chokri Thabet, Sihem Jebari, Ting Wang, Mathew Pryor, Surendra N. Kulshreshtha

First published in London, United Kingdom, 2019 by IntechOpen
IntechOpen is the global imprint of INTECHOPEN LIMITED, registered in England and Wales, registration number: 11086078, 7th floor, 10 Lower Thames Street, London, EC3R 6AF, United Kingdom
Printed in Croatia

British Library Cataloguing-in-Publication Data
A catalogue record for this book is available from the British Library

Additional hard and PDF copies can be obtained from orders@intechopen.com

Agricultural Economics - Current Issues
Edited by Surendra N. Kulshreshtha
p. cm.
Print ISBN 978-1-78984-049-0
Online ISBN 978-1-78984-050-6
eBook (PDF) ISBN 978-1-83962-637-1

For EU product safety concerns:
IN TECH d.o.o., Prolaz Marije Krucifikse Kozulić 3, 51000 Rijeka, Croatia, info@intechopen.com or visit our website at intechopen.com.

Meet the editor

Surendra N. Kulshreshtha is currently a professor in the Department of Agricultural and Resource Economics at the University of Saskatchewan, Saskatoon. He received his PhD in Agricultural Economics from the University of Manitoba. He was a visiting scientist at the International Institute for Applied Systems Analysis at Laxenburg, Austria. He has also participated in several oversees projects in Indonesia, Zambia, and India through the Canadian International Development Agency. His past research has been in the areas of commodity markets, econometric models, economic impact analysis models, greenhouse gas emissions, climate change, and impacts on agriculture, including extreme events (droughts) in the Canadian Prairie Provinces. Based on his contributions, in 2004 the Canadian Agricultural Economics Society selected him as a Fellow of the Society.

Contents

Preface

Agricultural economics is a subject area that contains many topics of interest to society. Originally focusing on farm- and rural-level issues, over a period of time it has grown to include many facets that touch human life. This book provides a glimpse of some of these issues facing society today. It is divided into four sections. The first section contains a chapter on the genesis of agricultural economics. The field of agricultural economics started as farm-level economics; however, over time, it has added many other related areas that are relevant to humanity. Its methodology has also become more complex, from the simple cost accounting framework used initially to complex modeling now. The second section is devoted to marketing issues. Here we have one paper (Chapter 2) dealing with value chain analysis for dates in the Gulf Cooperation Countries. The third section is devoted to resource economics through two papers. The first (Chapter 3) is on water scarcity in Tunisia and its impact on Tunisian agriculture. In addition, adaptations undertaken to reduce its negative impacts are also described. The second (Chapter 4) looks at crop conservation for onions and potatoes in Burkina Faso. The last section is devoted to environment-related issues facing agriculture. The first paper in this section (Chapter 5) is on the effect of value chain-induced constraints on conservation agriculture. This type of agriculture has many environmentally friendly attributes but faces a number of constraints in its adoption. Climate change is addressed in this section (Chapter 6) by examining the impact of climate change-induced salinity in Bangladesh. This is accompanied by an assessment of local techniques for crop conservation. A growing response to global food security could be achieved through rooftop gardens, which is addressed in the last chapter (Chapter 7) as a case study for its use for Hong Kong. The willingness to pay for such activities is also examined here. The editor hopes that these contributions provide valuable information to decision makers in their respective jurisdictions.

Dr. Surendra N. Kulshreshtha
University of Saskatchewan,
College of Agriculture and Bioresources,
Saskatoon, Canada

Section 1

Introduction

Chapter 1

Introductory Chapter: Agricultural Economics

Surendra N. Kulshreshtha

This book was planned to undertake an assessment of some recent developments in the area of agricultural economics. Although all aspects of this subject matter were welcome, only a selected few could be included in it. The book, therefore, describes only some of the areas in which agricultural economists are engaged. Nonetheless, it still begs the question "What is agricultural economics or what do agricultural economists do?"

As we know, agriculture is an important industry in most countries, because of its ability to provide the most basic necessity for human existence—food. On account of sheer growing masses of people, especially in developing countries, demand for food would increase. Associated with this would be a change in composition of diet resulting from higher income levels in most countries. As a result, even at present, in many developing countries, agricultural employment dominates, in spite of the fact that many farm workers are migrating to the non-farm locations in search of a better life. However, as a country develops economically, the relative importance of agriculture declines [1]. The primary reason for that is the Engel's Law (suggested in the nineteenth century), which is based on the fact that as consumers' purchasing power (incomes) increases, the proportion of income spent on food declines. This has been witnessed in many developed nations over the last few decades. However, one should keep in mind that although direct employment in agriculture in these countries is low, agriculture is still an important industry. This is because many of them had attained their development goals through moving labor and capital from agriculture, making it a source of economic growth. Furthermore, its importance is also indicative of current food and nutrition issues facing many of the countries.

1. Brief genesis of agricultural economics

As clear from the title—agricultural economics, the discipline grew out of two major parent disciplines—agriculture (or agronomy) and economics (study of scarcity leading to decision making by humans through use of resources). It arose in the late nineteenth century, combining the theory of the firm with marketing and organization theory, and developed throughout the twentieth century largely as an empirical branch of general economics [2]. Its popularity in the U.S.A. may also be credited to the establishment of the United States Department of Agriculture (USDA), which arose as a result of importance of agriculture as well as by rich data collected by the USDA, beginning in the mid-nineteenth century (based on [2], p. 2). The first professional note of agricultural economics was in 1907 where the American Economic Association (AEA), in its annual meeting, devoted a session to "What is agricultural economics?," which became a regular topic at these meetings [2]. This perhaps resulted in the creation of the American Farm Economics

Association in 1919, which retained its name until 1968 when it became the American Agricultural Economics Association (AAEA). Similar developments took place in other parts of the world, although details on them are either not available or very hard to collect.

An earlier precursor of agricultural economics was farm management, which is a science of organizing and combining people, natural and material resources for the purpose of crop and livestock production in order to maximize profit while optimizing input use. It grew out of the discipline of agronomy, which is concerned about physical management in agriculture—mechanization, cultural practices, land use, and labor utilization [3]. However, there was a lack of financial considerations in management decision, which was filled through the application of economic laws and theorems. Inclusion of financial management resulted in the creation of the American Farm Management Association in 1910 [4]. However, most preoccupation of the farm management professionals was undertaking farm surveys to determine production costs among groups of similar farms to gain an appreciation of what type of farm was most successful. Heady [5] criticized this discipline as having a methodology that was rooted in descriptive positivism, empiricism, and inductive procedures. This led to the development of farm economics discipline that later on was known as the discipline of agricultural economics.

2. Changing scope of agricultural economics

Agricultural economics is an applied area as well as a hybrid discipline. The field of agricultural economics includes application of economic science tools to the agricultural sector [6]. In fact, it is a branch of applied economics that takes the tools of both micro and macroeconomics and uses them to solve problems in a specific area [7]. Manning [3] suggests that agricultural economics grew out of a merger between rural economy and farm management in the early part of the twentieth century.

Application of economic principles to agricultural pursuits is not new. In fact, before the study of farm management was very old, it began to employ principles, theory, and information from economics [8]. Some of these included neoclassical political economy and the theory of the firm applied to the farm production, credited to Marshall [9]. Agricultural economics discipline included many of these applications and further evolved over a period of time. The importance of agriculture in economic development has led to pressing demands on the role in which agricultural economists are required to play [10]. In fact, Gardner and Rausser [11] suggest that the subject matter of agricultural economics has both broadened and deepened in recent years. For example, since the agricultural sector has undergone constant changes under the impact of new technology, shifting demands, and evolving institutions, one of the major areas of activity in agricultural economics is the study of policy issues. Assessment of impact of a given policy on the welfare of various members of the society is a major preoccupation of agricultural economists. In addition to the development of policy analysis, other areas were also added to the scope of agricultural economics. Included in this context was the study of rural economies. Agriculture was credited for supporting the rural economy directly or indirectly, which brought the study of rural development within the purview of agricultural economics.

In addition to the above two areas (over a period of time), one has witnessed several other additions to the discipline. Among these are areas related to natural resources and economic development (particularly in the developing countries). Studies related to natural resources tend to look at the role of such resources in the production of economic goods, including food. Environmental issues facing the society led agricultural economists to engage in environmental policies related to agricultural production. In the field of economic development, evaluation of new projects for economic development, using tools such as economic impact analysis or benefit-cost analysis, is now included in the discipline. Thus, agricultural economists study the sometimes conflicting needs of farmers, wildlife species, foresters, and outdoor recreationalists [12].

Though the term "agribusiness" had not been used until more recently, agricultural economists had been making significant contributions on issues related to agribusiness for many years [13]. This area emphasizes an integrated view of the food system that extends from research and input supply though production, processing, and distribution to retail outlets and the consumer. This research evolved along two parallel levels of analysis: (1) the study of coordination between vertical and horizontal participants within the food chain, known as agribusiness economics, and (2) the study of decision-making within the alternative food chain governance structures, known as agribusiness management [14]. Related to these are the issues dealing with trade and development. In fact, trade and development have dominated agricultural economics research particularly after the 1980s, with a more recent addition of food consumption and supply chain analysis in the food industry using an industrial organizational approach [2].

There are many other areas that have been brought within the scope of agricultural economics, as evidenced by planning of several parallel sessions in a typical agricultural economics conference. As an example, the 2018 Annual Conference of the AAEA included: finance, risk analysis, environmental and resource economics, behavioral and experimental economics, rural development and regional economics, price analysis and econometrics, food and agriculture marketing and policy analysis, supply chain management, institutional economics, international trade and international development, natural resource economics (including land and water economics), production economics, plus many others, as sessions for presentation of current research. Many more areas can be listed in addition to these listed above to do full justice to what agricultural economists do.

3. Proposed objective of the book

The major objective of this book was to explore the current and future issues in agriculture from an economic point of view. In addition, as a guidance to policy-makers, it was hoped that this book can also provide information on the future shape of agriculture in the world. In many parts of the world, food production is not able to meet the demand for the growing population. In addition, increased income in many regions has changed the composition of the food basket demanded by today's consumers. In many nations, rural areas are losing people toward urban areas—a process called urbanization. This has created an additional stress on the economic health of agriculture. In the future, land degradation, climate change, air and water pollution, depletion of freshwater resources, loss of forested areas, and other similar trends would add further stress on the economics of the agriculture industry, and on the people associated with it. This book was intended to provide readers (policy-makers as well as academics) with a comprehensive overview of the current and future state of the art in assessing economics of agriculture and how it is going to change over time in various parts of the world. Although this was the initial intent of the book, its present contents are not totally reflective of all these areas. In fact, contents of this book relate to just a few of them that are of interest to agricultural economists currently.

Author details

Surendra N. Kulshreshtha
Department of Agricultural and Resource Economics, University of Saskatchewan, Saskatoon, SK, Canada

*Address all correspondence to: suren.kulshreshtha@usask.ca

References

[1] Johnson DG. Agricultural Economics. 2019. Available from: https://www.britannica.com/topic/agricultural-economics [Accessed: April 2, 2019]

[2] Runge CF. Agricultural Economics: A Brief Intellectual History. Working Paper WP06-1. St. Paul, MN: Center for International Food and Agricultural Policy, University of Minnesota; 2006

[3] Manning T. The roles of agricultural economics. Canadian Journal of Agricultural Economics. 1980;**28**(3):1-5

[4] Taylor HC. The history of the development of the farm economic association. Journal of Farm Economics. 1922;**4**(2):92-100

[5] Heady EO. Elementary models in farm production economics methodology. Journal of Farm Economics. 1948;**30**:201-225

[6] Seyla T. Agricultural Economics. 2019. Available from: https://studylib.net/doc/8888806/agricultural-economics [Accessed: April 2, 2019]

[7] Morgan W. Agricultural Economics. 2019. Available from: https://www.google.co.uk/search?ie=ISO-8859-1&oe=ISO-8859-1&q=Agricultural+Economics&btnG=Search&domains=www.studyingeconomics.ac.uk&sitesearch=studyingeconomics.ac.uk&gws_rd=ssl [Accessed: March 27, 2019]

[8] Jensen HR. Farm management and production economics, 1946-70. In: Martin LR, editor. A Survey of Agricultural Economics Literature. Vol. 1. Minneapolis: University of Minnesota Press; 1977. pp. 3-92

[9] Marshall A. Principles of Economics. 1st ed. London: MacMillan and Co; 1890

[10] Fenyes TI, van Rooyen CJ. The contribution of agricultural economists to development. Journal of Development South Africa. 1985;**2**(2):271-286

[11] Gardner BL, Rausser GC. Handbook of Agricultural Economics. Vol. 1A, B, 2A, B. Amsterdam: Elsevier; 2002

[12] Moncrieff PM. Agricultural Economics. 2012. Available from: https://www.thecanadianencyclopedia.ca/en/article/agricultural-economics [Accessed: April 1, 2019]

[13] King RP, Boehlje M, Cook ML, Sonka ST. Agribusiness economics and management. American Journal of Agricultural Economics. 2010;**92**(2):554-570

[14] Cook ML, Chaddad FR. Agroindustrialization of the global agrifood economy: Bridging development economics and agribusiness research. Agricultural Economics. 2000;**23**:207-218

Section 2

Marketing

Chapter 2

Date Palm Value Chain Analysis and Marketing Opportunities for the Gulf Cooperation Council (GCC) Countries

Boubaker Dhehibi, Mohamed Ben Salah and Aymen Frija

Abstract

In order to develop a sustainable date palm production system in the Gulf Cooperation Council (GCC) countries of the *Arabian Peninsula,* an analysis of the date value chain in these countries was undertaken. Through the mapping of the chain, the overall objective was to identify the processes where values are created and how they are distributed among stakeholders along the entire date palm value chain. The method used in this analysis was based on an assessment of the data gathered from the multi-stakeholder surveys implemented in the three case studies of the GCC countries: Kingdom of Saudi Arabia (KSA), Oman, and Kuwait. The empirical findings reveal several problems and constraints that might affect the future of the GCC date palm sector. Therefore, development of a competitive supply date palm chain (both market and agribusiness development) could provide a greater contribution to the GCC economy if producers paid more attention to marketing of this very important food commodity. Hence, date palm production is no longer a way of life but nowadays is considered as an investment option and source of revenues for many stakeholders if the GCC region.

Keywords: value chain, date palm, marketing strategy, GCC countries

1. Introduction

Date palm cultivation in the Gulf Cooperation Council (Kingdom of Saudi Arabia, Qatar, Bahrain, Sultanate of Oman, United Arab Emirates, and Kingdom of Bahrain) has a long history, yet the efforts exerted by the individual countries on its research and development, although significant, are still insufficient and fall below expectations. In general, the product quality is still low, the field and post-harvest losses are high, and the date products and by-products utilization needs improvement. Therefore, the status of date palm cultivation in the GCC countries and the enhancement of quality of produce cannot be overemphasized. To address the above mentioned constrains, the GCC countries ranked date palm as one of the high research priority as reflected in priority setting for agricultural research in the Central and West Asia and North Africa (CWANA) region [1].

In this regards, the project "*Development of sustainable date palm production systems in the GCC countries of the Arabian Peninsula*", funded by the GCC Secretariat, was implemented, in partnership, by various ministries of agriculture, agricultural authorities, and agricultural research institutions and universities in the six GCC countries of the Arabian Peninsula (Kingdom of Bahrain, United Arab Emirates, State of Kuwait, State of Qatar, Sultanate of Oman, and Kingdom of Saudi Arabia - KSA) and the International Center for Agricultural Research in the Dry Areas (ICARDA). The major objectives of the project are to improve date palm productivity per unit of water and rationalize the use of the available resources in order to make production sustainable.

2. Objectives of the study

This chapter provides a description of the actual situation for the date palm value chain in the GCC countries, while providing strategic short term perspectives for a more effective and inclusive date palm market sector, constraints and challenges that have to be taken into account as well as proposed strategies to enhance systemic changes in the sector necessary to progress towards more solid and sustainable date palm value chain in the GCC. A specific attention is given to the challenges and constraints in the date palm sector in the GCC countries.

Through the mapping of the chain, the overall objective of this study is to identify the processes where values are created and how they are distributed among stakeholders along the date palm value chain. A special focus is devoted to policy instruments used by the Government to solve specific problems characterizing the chain and consequently enhance the development of the sector and see in particular to what extent those instruments are value creating.

With this view in mind, the present study has been designed with the following specific objectives:

- To overview the date palm sector in GCC countries.
- To analyze the existing value chain of date palm marketing with special attention to the international date palm markets and opportunities for GCC countries.
- To provide strategies and interventions for the GCC date value chain with potential for significant development of value-added.

3. Overview of the date palm sector in the GCC

3.1 Date palm production

Date palm production is a strategic sector in most of the Arab countries including the GCC ones. The sector is one of the oldest economic activity in the Arabian Peninsula and continue to play a key role in the welfare, culture, history, environment, and nutrition of its population. At present (in 2016), the Arab Region is the world leader of date cultivation with almost 75% of global area under date palm, around 77% of world production and approximately 69% of world total export of dates. In addition to the importance of dates for domestic consumption, this sector is also a source of employment, income generation, and trade in many of these countries. In some very arid areas, date fruit remains as an important source of subsistence and resilience for local population, given its adaptability to harsh environment and tolerance to high temperature, salinity, drought and other severe arid conditions.

According to the FAO statistical database [2], the GCC countries such as KSA, Oman, and United Arab Emirates (UAE) have the highest harvested areas in 2016 with respectively 145,516 ha; 24,120 ha and 93,561 ha in the three countries. While this area was increased in KSA during the last two decades (from 142,450 ha in 2000 to 145,516 ha in 2016), it has rather been decreasing in both Oman (from 35,508 ha in 2000 to 24,120 ha in 2016) and UAE (from 185,330 ha in 2000 to 93,561 ha in 2016). This decrease was the highest in Oman with around 49.50% between 2000 and 2016. This This decline is mainly attributed to a combination of various factors including increased soil salinity in major date palm-growing regions, desertification in areas adjacent to the desert in central Oman, heavy insect pest infestation such as dubas bug and red palm weevil, and urbanization of rural areas.

With respect to pace and trends in planted area, production and productivity varied considerably between the GCC countries although perhaps it was a result of the special attention paid and considerable government support during the last few decades. The harvested areas in Bahrain, Kuwait and Qatar are still very limited with respective values of 3986 ha; 3021 ha; and 2407 ha during 2016. In these three countries, the planted area has been quickly progressing during the last two decades with an average annual increase of about 20, 40, and 1.38% in respectively, Bahrain, Kuwait, and Qatar. The highest average yields (calculated over the period 2000–2016) are recorded in Kuwait, Qatar, and Oman with respectively 22.03; 11.13; and 10.34 tons/ha.

The assessment of the date palm yields reveals that the average yield at the global level is around 6 tons/ha. In the GCC countries, the lowest average yields are in UAE, KSA, and Bahrain, with respectively 5.89, 6.38, and 7.56 tons/ha in the three countries. These yield values combined to the statistics on harvested areas makes Saudi Arabia the top producer of dates in the GCC region, with an average annual production[1] of 885,542.8 tons/year, followed by UAE producing and average of 667,569.8 tons/year. It is clear that the productivity of the date palm tree varies within individual countries, and mainly depends on the agro-ecological systems, variety and farming system adopted.

3.2 Marketing systems

3.2.1 Domestic consumption

In terms of consumption, GCC countries vary widely in their per capita date consumption. According to Frija et al. [3], date consumption per capita is highest in Oman with a value of 68 kg/capita/year, followed by Saudi Arabia with a value of 34 kg/capita/year. Sultanate of Oman greatly outpaces all GCC countries in per capita date consumption, which is more than double that of the KSA, the second highest ranking in date consumption among Arab and GCC Countries. The per capita date consumption for the other countries is comparatively low. The trends of date consumption per capita in the study countries reveals, in addition to being low, that it is also decreasing in most of the GCC countries over the period 2000–2013.

[1] Calculated over the period 2000–2016.

3.2.2 Livestock feeding use

In addition to human consumption, date pits and dates falling down from palms before maturity are used as animal feed. There is some use for feed of final product in some countries. This practice is not only observed in GCC countries but is also frequent in other North African countries, such as Tunisia. Some portion of date production is date wasted. Recent statistics from FAO (several years) indicates that up to 38% of the date production was fed to animals in UAE during 2013. For Oman, this figure is around 4%. The volume of wasted dates is also important in the study countries. The lowest wasted percentage (of production) is recorded in KSA (1%), while a highest rate of 13% is recorded in Kuwait.

3.2.3 Domestic market and structure

In the majority of GCC countries, date products' marketing is dominated by local marketing process since export marketing, with the exception of UAE and KSA, is still a small percentage of local production. Domestic marketing of dates is free of direct government involvement and is sole responsibility of the private sector and to a lesser extent of the producers. It takes many forms. At the domestic market, date marketing is passing through two avenues:

- Direct traditional marketing to consumers: Where the producers sell their dates after harvesting directly to the local markets localized in the production areas and the neighboring markets without any further processing. The dates are commercialized without proceeding to any process, such as sorting, grading, steaming, and washing. The dates are marketed under two stages according to the maturity of the fruit: The stage of secret and wet (fresh) and the final matured fruit.

- Marketing to the date palm factories: In general, the dates delivered to the factories are of the best quality as the date palm producers deliver their products to the local existing factories under pre-fixed norms and standards fixed by the factory. The price is fixed on the basis of quality of the fruits and the supplied quantities.

3.3 Marketing channels of dates

Food marketing, according to Kohls and Uhl [4], is "the performance of all business activities involved in the flow of food products and services from the point of production until they are in the hands of consumers." For dates, these activities include harvesting, processing, packaging, and transportation/shipment to local or export markets. A marketing channel describes the movement of a product or commodity from the site of production to the place of consumption. It may include transportation, handling and storage, ownership transfers, processing, and distribution. The marketing channel for dates includes initial processing at farm level after harvest, transport to the local market directly (especially for the highly perishable fruits) or to the packing plant, processing and packaging at factory level and transport to the final consumers. There are no universal set of marketing channels because each country is unique, and institutions involved operate differently under different sets of regulations. Therefore, in the GCC countries, the differences in the marketing channels are minima's, and include on-farm selling, retailers, local markets, date factories, and consumers markets.

3.3.1 Example of Kingdom of Saudi Arabia marketing channel for dates

The KSA is an important stage in both the traditional and modern methods of marketing. There is packing for long-distance transportation (local and export) and packing for the final consumers at supermarkets. Dates can be sold at the farm gate and from there dates are marketed either directly to the final consumer at the local market or to the local wholesale market. In practice, four existing/dominant channels for the commercialization of dates are dominant is KSA and are presented as follows:

1. Producer/farmer → Consumer
2. Producer/farmer → Retail Trader → Consumer
3. Producer/farmer → Date palm factory → Retail trader → Consumer
4. Producer/farmer → Wholesale trader → Date palm factory → Retail Trader Consumer

3.3.2 Example of Oman marketing channel for dates

In Oman, the marketing channels for dates include on-farm selling, retailers, local markets, date factories, and export. There are therefore mainly three channels through which date flow from the farm to local and foreign consumers/export market:

- Dates can be sold at the farm gate and from there dates are marketed either directly to the final consumer at the local market or to the local wholesale market.
- Dates can be marketed directly to wholesalers and from there to retailers either before reaching the local market or directly to the local market.
- Dates can be marketed directly to wholesalers and from there to the processing factories for processing and packaging before being shipped to the retailers' trader and then to the final consumer (local or export).

3.4 Exports market channels

The trade matrix of dates for the GCC countries is represented through a list of exported quantities from each of these countries to different destinations in the world. However, due to the high number of destinations, **Table 1** summarized only

Countries	2012			2016		
	Number of countries	Exported quantities (tons)	Value of export (1000 US$)	Number of countries	Exported quantities (tons)	Value of export (1000 US$)
Oman	23	5814.9	7745.6	44	15,699.836	12,650.393
Bahrain	2	43.8	33.1	3	16.716	2868
Kuwait	19	363.9	337.7	14	405.777	542.958
Qatar	na	na	na	na	na	na
KSA	60	64,299.0	74,859.5	64	117,017.911	141,564.174
UAE	98	304,090.5	129,177.2	110	275,862.901	160,215.460

Source: UN COMTRADE data base; (na: not available). Number of countries shows the number of countries to which each of the respective GCC countries is exporting dates; Exported quantities are expressed in tons; value of export is expressed in 1000 US$.

Table 1.
Number of dates exporting markets for each of the GCC countries.

the number of countries to which each of the respective GCC countries are exporting to. Results outlined in **Table 1** reveals that KSA and UAE are the most active in terms of market diversification expressed by the number of countries to which they are exporting dates. In 2016, UAE was exporting to 110 countries, while Saudi Arabia was exporting to 64 countries.

The analysis of the intensity of date imports-exports, particularly among the GCC countries (**Figure 1**) based on the availability of recent data, shows again that KSA and UAE are the most active in terms of date exports to the different GCC countries. On the GCC market, UAE is mostly exporting to Oman, followed by KSA and Qatar. KSA is mostly exporting to UAE followed by Kuwait and Qatar. Most of the Omani dates is also exported in the destination of UAE, which is showing that UAE is the biggest exporter and importer partner of Oman.

In the export channel, dates are sold directly to the processing factories for processing and packaging before being shipped to export markets. It is indicated that this channel is taking a considerable share in some countries, such as KSA and UAE, in comparison to the rest of GCC countries. The dates are produced, harvested, sorted, graded, processed, packaged, and transported in an efficient, safe, and with high quality management. This process determines the final market value of the dates, as shown in **Figure 2**.

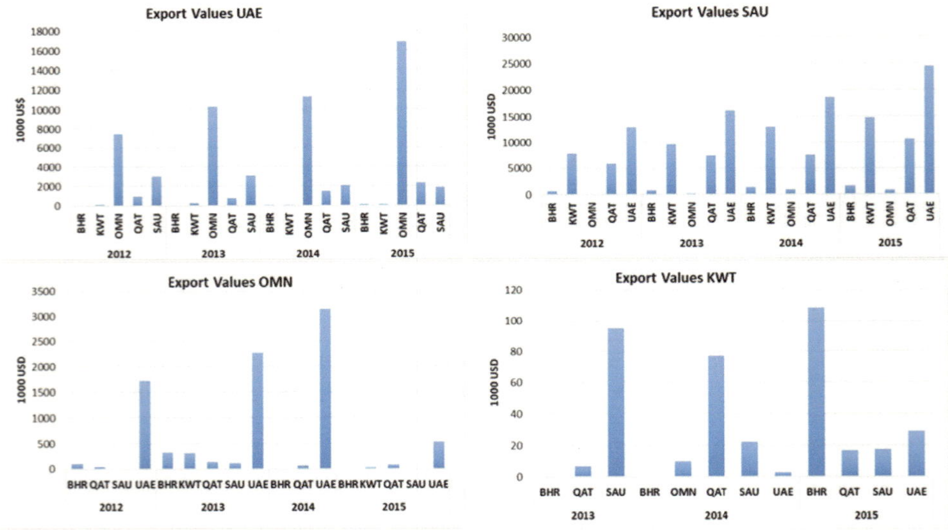

Figure 1.
Export trade of dates among the GCC countries (UAE: United Arab Emirates, SAU: Saudi Arabia; OMN: Oman; and KWT: Kuwait). Source: Own elaboration from COMATRADE UN database [5].

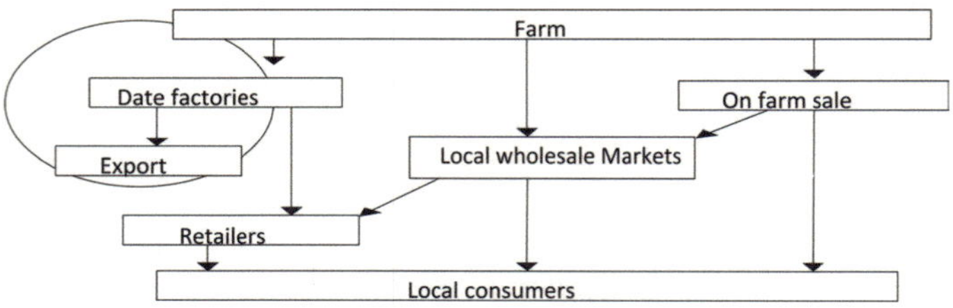

Figure 2.
Omani date marketing channels with date export supply chain indicated by circle. Source: Mbaga [6].

4. International date palm markets and opportunities for GCC countries

The GCC is considered as an important component of the regional and international date trading; the trend of the international marketing of dates shows clearly the weight of the quantity of date exported by GCC countries in the international market [3]. The fluctuation of the trend of the dates trading in the GCC countries is mainly due to the instability of volumes exported by the UAE and KSA. The analysis of date trade reveals that in 2016, GCC and North African countries are capturing more than 70% of the international market of dates from the Arab countries. **Figure 3** shows how this market is shared among the GCC countries and their direct competitors from North Africa, such as Tunisia, Algeria, and Egypt. In addition to Israel, who constantly holds more than 10% of the international market of dates, Tunisia is dominating in terms of market share, with an average value of around 20% over the last decade. Algeria and Egypt are simultaneously holding around 3.6 and 3.8% of the market.

As shown in **Figure 4**, the international market share of GCC countries in 2016 was about 28.26%. In terms of growth, it was clear that all GCC countries, including the least present on the international market are progressing quite positively with increasing shares from 1 year to another Such result confirm that these countries

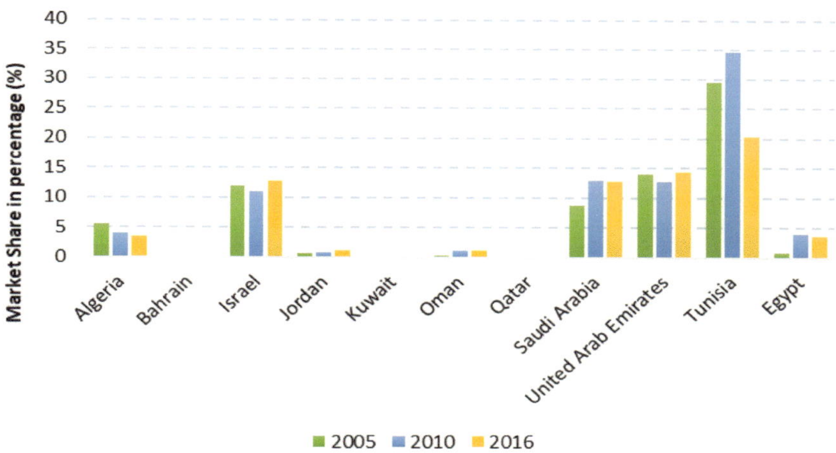

Figure 3.
Market share of selected dates exporting countries, including GCC countries and their competitors during the period 2005–2016. Source: Own elaboration from FAOSTAT (2018).

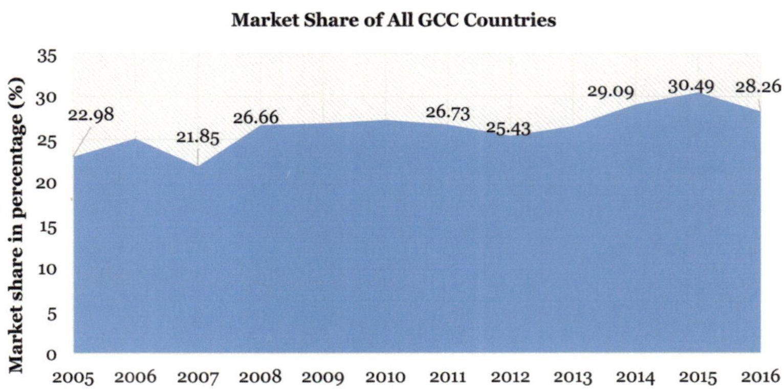

Figure 4.
Trend of the aggregated GCC market share (sum of Bahrain, Kuwait, Oman, Qatar, Saudi Arabia, and United Arab Emirates) during the period 2005–2016. Source: Own elaboration from FAOSTAT (2018).

together have strong potential for dominating the international dates market if additional efforts (concentrating on their dates international markets) are made mainly in the important as well as growing date importing countries.

In addition to the above, this share of the GCC countries has been increasing over the period from 2005 to 2016, from a value of 22.98% in 2005 to 28.26% in 2016. This change especially refers to the rapid increase of the shares of KSA, UAE, and Oman, as shown in **Figure 5**.

In terms of progress, all GCC countries, including the smallest share countries present on the international market are progressing quite positively with increasing shares over the period 2005–2016. This is especially true for Kuwait and UAE. Oman and KSA, also have the same trend with market shares increasing respectively from 0.34 and 8.64% in 2005 to 1.13 and 12.70% in 2016.

During this period (2005–2016), the gap between GCC and its competitors on the international market is progressively narrowing, mainly with North African countries (Tunisia and Algeria), Iran, Pakistan and Israel. This might be due to the important increase in the demand for dates at the international level through expanding to new markets (Asia and Africa).

Although date trading share of GCC is increasing in the global market, they are still facing an important competition from the countries outlined above (Tunisia, Algeria, Iran, Pakistan, and Israel). To deal with this increasing competition, regional cooperation and partnerships among these countries should be reinforced, given that their challenges, constraints and problems on facing the global date market are similar, and interventions to address and resolve these complex and interrelated

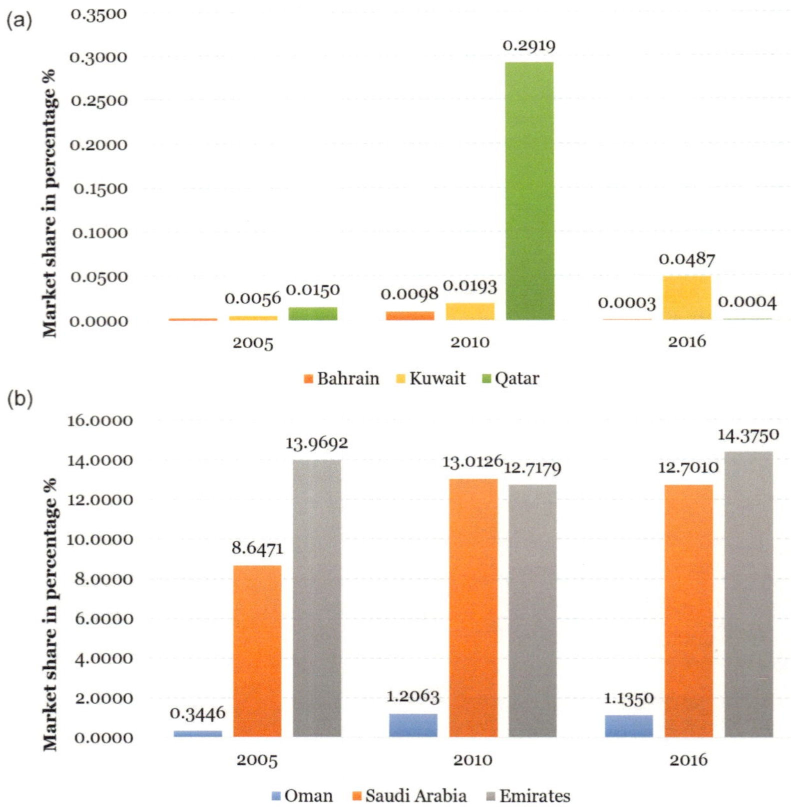

Figure 5.
Market share of GCC countries on the international date market: (a) Bahrain, Kuwait and Qatar; (b) Oman, Saudi Arabia, and Emirates. Source: Own elaboration from FAOSTAT (2018).

challenges and problems are difficult for a single country to solve on its own. Thus, coordination between the different trade strategies of the GCC countries, through specialization, division of tasks, and strengthened coordination can generate important opportunities for gaining and sustaining their share on the world market of dates.

5. Date value chain strategy in the GCC countries

5.1 Data sources and data collection

The data for this study have been collected through semi-structured questionnaires targeting the main actors in the date palm sector. Data were gathered for three countries: Saudi Arabia, Oman and Kuwait. **Table 2** illustrate the number of surveys conducted in each country mentioned above.

5.2 Methodological framework

The method used in this analysis is an assessment of the data gathered from two different sources.

First, an exhaustive literature review from various types of documents, both published and unpublished, related to date palm value chain in the GCC countries. Documents includes scientific reports, books, journal articles, working papers, research reports, web-based publications (including national and international databases), workshop proceedings, national policy documents, programme frameworks, etc.

Second, a participatory and consultative approach comprising multi-stakeholder consultations and surveys implemented in three potential GCC countries: Saudi

Nature of the survey	KSA (number of interviewed stakeholders)	Oman (number of interviewed stakeholders)	Kuwait (number of interviewed stakeholders)
Date palm growers	40	38	62
Date palm consumers	09	—	11
Date palm retailers	08	04	10
Date palm wholesale traders	08		
Date palm processors	08	—	—
Date palm transporters	08	14	—
Date palm importers (production inputs)	09	18	—
Decision makers	07	—	—
Researchers and extension staff	10	47	—
Date palm input providers	20	—	—
Total interviewed stakeholders	**127**	**121**	**83**

Source: Own elaboration form surveys implemented in the GCC countries (2017).

Table 2.
Nature and number of surveys conducted in each country.

Items	Assessment indicators
Fresh dates for local consumption	High in all GCC countries
Fresh dates for exports	Almost high in KSA and UAE
Industrial dates for exports	Low in Oman, Kuwait, and Bahrain
Industrial dates for domestic consumption	Low in Oman, Kuwait, and Bahrain
Animal feeding dates	Important in Oman, Kuwait and UAE
Utilization of by-products	Low in all countries
Farm management	Low in all countries

Source: Own evaluation from survey database (2017).

Table 3.
Date palm supply value chain synthesis.

Arabia, Oman, and Kuwait. The surveys targeted the following date palm stakeholders: growers, consumers, retailers, wholesale traders, processors, transporters, importers, decision makers, research and extension staff, and input providers. The method was based on a synthesis summarizing the key messages issuing from each one of the indicated actors. The framework consisted of the following two processes. The first ones focused on data collection from surveys through involving the main value chain stakeholders. The second one focused on the validation of the identified results with special attention to the key development strategies and interventions.

This process, which was developed and implemented by a large number of actors actively involved in the three selected countries date sector (**Table 3**), led to three development strategies: (i) Date palm supply chain development and management; (ii) market and marketing development, and (iii) agribusiness development.

5.3 Challenges and constraints in the date palm value chain in the GCC countries

The analysis of the data collected reveal several problems and constraints that might affect the future of the GCC date palm sector. Such problems include:

- Low quality varieties, low product quality and consequently low returns.
- Poor farm Management (Inefficient use of date products and by-products).
- Pests and diseases and inadequate Integrated Pest Management (IPM) control.
- Harvesting, processing and marketing (high post-harvest losses)
- Lagging processing sector.
- Low quantity of exports.
- Shortage in national qualified and trained staff & labors.
- Insufficient research and development activities.

The assessment of the date palm supply value chain, on the basis of the data collected from the different actors, allowed us to identify the following indicators related to each GCC country (**Table 3**).

According to **Table 3**, the date palm supply value chain assessment indicates a high level in local consumption of fresh dates although per capita consumption is declining in most of the GCC countries because of the new habits dominating the new youth generation. The industrial dates for domestic consumption remain low mainly in Oman, Bahrain and Kuwait given the consumers preferences for the GCC countries in general and these countries in particular is more towards fresh dates. With respect to trade, KAS and UAE are the leading date exporting countries both for industrial and fresh dates. Oman, Kuwait, and Bahrain are the lowest exporting countries compared to their production (i.e., Oman).

Findings reveal also that part of date production is used as feed, mainly in Oman, Kuwait and UAE. Up to 30% of production was served as feed to animals in UAE during 2013 [3]. In addition, most of the date by-products are produced in traditional manner and mainly limited to the production of date paste, Jam, and date syrup. The introduction of new potential uses of dates and date products is modest, and consequently the utilization of by-products is low in all GCC countries.

Finally, the results of date palm supply chain assessment indicate a lack of an improved and advanced management and lack of knowledge of good agricultural practices applied to palm date farming systems in almost all GCC countries.

5.4 Date value chain development strategy in the GCC countries

5.4.1 Farming practices: farmer education and farming management practices

Improved production and the whole production farming system is an essential part of any successful agribusiness program for the date palm sector in the GCC countries. The main challenges facing the production of date palm mainly relate to the weakness of the productive capacity of small farmers due to the lack of good pollen and low productivity of cultivated varieties compared to other varieties.

The lack and the reduction of the skilled labor for date palm practices affect tremendously the production and the quality of the dates. Therefore, a holistic extension system is more than needed, mainly for the non-commercial farms, to provide training, advices, monitoring, and know-how for the date farming system. Staff should be trained in processing techniques to improve efficiency and quality and to produce on a larger scale. In addition, staff needs to be capacitated in post-processing quality assurance and control, better marketing and accessibility to market and improving quality standards and quality extension certification. In addition, capacity building of extension institutions, including the staff of local extension departments is more than needed. Capacities of such staff need to be built to better (more effectively, larger coverage, more permanent support) support producers and promote good agricultural management, pest and disease management using participatory training approaches (learning by doing and action field schools). This program should provide affordable and simple small-scale tools for the key tasks in this farming system (pollination, irrigation, fertilization, harvest, and post-harvest techniques). Such program could start at small scale level and later extended at large scale.

5.4.2 Variety selection

In the GCC countries, farmers grow more than 600 varieties of dates. Some varieties, given their own special set of characteristics, such as size, shape, color, and skin texture, are well appreciated by the consumers (at the national and international markets). Even if it is considered as an advantage for the date palm growers, in general, this variability creates obstacles in the sector, particularly during the harvest and post-harvest activities. This leads to many problems in processing and marketing (grading, sorting, quality, etc.) given the non-uniformity of such products. GCC producers are mainly recommended to focus on producing the high added values varieties to meet national and international market demands. In addition, date palm producers are advised to improve the date palm productivity through the cultivation of new higher yielding varieties.

GCC countries should produce the most important and commercial varieties. These as an example include: *Mejdoul* in KSA, *Fard* and *Khalas* in Oman, *Barhi*, *Sukkari*, in Kuwait. Nowadays, international markets are very demanding both for the industrial sectors and for the final users (consumers) such standards and norms. *Mejdoul* is at the top of the list, followed by *Khalas*, *Fard.* Since GCC countries have abundant production of these varieties, an effective effort should be made by all the actors to propagate them within the Gulf region. This effort should mainly focus on the producers as they are the ones who influence the future plantation. The decisions makers are also requested to play a major role in this strategy by providing support and enhancing the propagation of plantation of these varieties. Finally, these varieties, and other potential ones, can be also produced to satisfy the local market needs from these products. This could be enhanced through the facilitation of the establishment, in each country and even at the most regional level, of a national palm plant nursery, a pollen collection center, and a tissue culture laboratory in coordination with the Ministry of Agriculture and research institutions to develop better, shorter and productive varieties. In addition, this laboratory can also be building solid relationships with input/service providers to supply red palm weevil detection devices, to introduce automated pollination mechanisms, and industrial maturation equipment.

5.4.3 Post-harvest handling improvements

The biggest challenge for the date palm sector in the GCC countries is the large volume of dates that arrive at the same time on the market (which is around harvest time), resulting in low prices. Indeed, to have a good and decent market price, the date value chain should be strong. This is possible when the post-harvest handling capabilities: post-harvest handling of fruit is efficient. According to the assessment we made, in the majority of the GCC countries, the post-harvest handling is considered as the weakest node of the date palm value chain where greatest loss occurs.

Post-harvest handling of dates currently in some GCC countries (Oman) is generally very poor. The majority of small scale date farmers are absent and foreigner laborers are managing the farming system. These "new managers" are with minimal education, who must be taught proper post-harvest techniques from beginning to the end. Thus, farmer education to these laborers should be a long-term program requiring constant training, oversight, and compliance. Efforts on monitoring control by the owners, and funding for improving farming methods and equipment and building a processing plant to produce product meeting international standards will prove a major disappointment if owner's farmer's monitoring, control and investment in best practices is not seriously and continuously addressed.

This is will influence, some posteriori, the date industry that will not move up the value chain if the fruit is not properly handled from harvest to final processing, and consequently there is little that can be done in terms of processing if fruit arrives at the processing plant infested with insects and full of dirt and sand. Such fruit cannot be processed into an added value or commercial product to meet the minimum marketing requirements.

5.4.4 Processing and value added for date products

There is no doubt that value added products for date palm sector is the future of this industry in the GCC. Low quality dates are mostly absorbed by the local market at very low price for the consumers. There is also the use of the very low quality as feed for the livestock sector. The local markets already absorb all of the low-quality dates at a low price. Therefore, for the high date palm quality, a high quantity of waste is generated at various stages of date fruit value—chain/marketing. This date solid waste could be valorized from different perspectives and for several uses (pharmaceutical industries, confectionaries, handcrafts and furniture, etc.).

These suggested value-added products (date palm by-products such as leaves and wood for furniture and handicrafts, sweet sap, date sugar, date palm-based confectionaries products, etc.) should be produced in modern, efficient processing facilities capable of making a wide variety of industrial grade and retail products demanded by the local and the international markets. This is possible if the facilities are installed close to the production areas, regrouping farmers around agricultural cooperatives, which offer opportunities that date palm small holders could not achieve individually, such as providing good quality of (as fruits and, consequently a strong bargaining power and resource share that leads to win-win situation.

A systemic change that encompass the independence of external inputs, that promote better palm tree management and harvesting, improve storage, processing and packaging and work on creating higher value products for the local and international markets, is encouraged. This systemic change should be started by facilitating the linkages between decision makers (ministries) and research institutions, with producers and other key date palm chain actors, through promoting the use of applied scientific research methods to improve processing quality and develop new processing and storage techniques. There is also a need to build a trustable relationship with input/service providers to introduce new processing and packaging technologies and techniques.

5.4.5 Markets and marketing improvement

5.4.5.1 National markets

Several constraints and challenges are facing the local marketing channels of the date palm fruits. Such challenges are decreasing demand for date fruits, weak post-harvesting transactions, lack of appropriate and attracting boxes, no appropriate existing storage system till the marketing of fruits, lack of by-products markets, use of unimproved technologies for fruit packaging, no linkage between all key processing date palm actors, lack of expertise and technical staff for date palm manufacturing industries, rural-urban migration, urbanization, no feasibility studies regarding the profitability of the manufacturing industries of date palm products, and reluctance of investors to invest in the date palm sector. To overcome these constraints and to improve the marketing at the local level, it is crucial to provide the pertinent, complete and in time marketing information to date palm producers; activate the role of the marketing cooperatives; and development of coordination committees, associations, cooperatives, public and private companies for producers, manufacturers, and exporters of dates to prevent duplication and waste resources. Another strategy could be by including date fruits by the Ministry of Health, Education in the hospital, schools, etc. Furthermore, initiating local consumer's campaign, which will be set up to promote a "buy-local" products, would enhance consumer trust and pertinence in/to local produce. Government institutions, private sector organizations, and the Chamber of Commerce and other stakeholders might be encouraged to launch targeted marketing campaigns linked to new potential consumers/customers [3].

5.4.5.2 International markets

As it was stated above, the international markets are becoming very demanding in terms of standards and norms, mainly for the perishable products such as dates (as fruits) This will be a challenge for local producers to be competitive in the international markets. In 2015, since about 30% was the share of the GCC countries in the international date market, there is a strong potential for dominating the international date market. This is feasible if there is a strong coordination on trade strategies between the GCC countries in terms of providing safe and secure products; adopting standard specifications and norms in the producing countries to reach the world market with distinct varieties reflecting the position of these countries in the production of date fruits; conducting studies to understand the international markets (volume, size, main competitors, marketing strategies, etc.), their needs, laws, regulations, and providing the information on the national key actors (producers, manufacturers, exporters, etc.), supporting and encouraging the participation of producing companies in the international trade fairs and with supporting exports policies, and finally conducting campaign on advertising and promotion of dates and its nutritional and healthy benefits through international meetings, conferences, festivals, and Attaché offices.

6. Concluding remarks, policy implications, and recommendations

This study analyzed the value chain of date palm fruit in the GCC Countries. The value chain development, sustainability and competitiveness are very crucial issues facing the agribusiness sector of the GCC economy, particularly for the small holder's growers. The GCC agribusiness is been noted for the sale of raw agricultural products palm fruits is one of the major cash crops in the majority of these countries. Dates could provide a greater contribution to the GCC economy if producers paid more attention to its production and marketing. Date palm production is no longer a way of life but nowadays is considered as an investment option and source of revenues for many stakeholders.

The assessment of the supply date palm value chain in the GCC countries suggests the following:

- Reduction in the consumption of dates.
- Decreasing trends on the exports of fresh dates.
- A low consumption level for the industrial dates.
- An important level on dates used for animal feeding.
- A low level of utilization of date by-products.

Moreover, several constraints and challenges affecting the date palm value chain in the GCC including are revealed and are summarized as follows:

- Low quality varieties, low product quality, low exported quantity, and consequently low returns.
- Poor farming management (Inefficient use of date products and by-products).

- Pests and diseases and inadequate integrated pest management (IPM) control.
- Harvesting, processing and marketing.
- High post-harvest losses.
- Lack of efficient processing system (storage, transportation, etc.).
- Inefficient use of dates products and by-products.
- Low exported quantity and low competitiveness of processed date palm products on the international markets.
- Shortage in national qualified and trained staff & labors.
- Insufficient research and development activities in the date palm sector.

Therefore, to enhance the growth of the date palm value chain in the GCC countries, three development strategies are to be considered towards a dynamic and systemic change:

- Supply chain development and management
    - Improving date palm farming practices and initiation of better post-harvest handling procedures;
    - Processing and production of value-added products by focusing on added value varieties;
    - Export premium date products with consistency in supply of high-quality varieties.
- Market and marketing development
    - Establishing a basic data on the local and export date marketing sector;
    - Improve the efficiency of the marketing process to guarantee good prices to the farmers and to decrease the final market prices by limiting the number of intermediate operators;
    - Producing in-demand products at competitive prices;
    - A Pre-harvest Best Management Practices (BMPs) will certainly ensure that the quality standards for export are met.
- Agri-business development
    - Introducing of new potential uses of dates and date products in the pharmaceutical industries, confectionaries, agro-food industries, handcrafts and furniture,
    - Support the investment for the Small Business Enterprises (SBE's);

- Promoting the use of applied scientific research methods to improve processing quality and develop new processing and storage techniques.

Acknowledgements

We would like to express our sincere gratitude and appreciation to the Gulf Cooperation Council (GCC) Secretariat for funding this research conducted in the framework of the "*Development of sustainable date palm production systems in the GCC countries of the Arabian Peninsula*" project.

We are very grateful to the Ministries of Agriculture, Agricultural Authorities, and Agricultural Research Institutions and Universities in the GCC countries of the Arabian Peninsula for their continuous support and great collaboration in the implementation of the project activities.

Special thanks go to the all stakeholders (date palm growers, consumers, wholesale retailers, decision makers, date palm processors, date palm transporters, date palm importers for production inputs, researchers and extension staff) in KSA, Oman, and Kuwait for their great collaboration during the data collection process.

Author details

Boubaker Dhehibi[1*], Mohamed Ben Salah[2] and Aymen Frija[1,3]

1 Resilient Agricultural Livelihood Systems Research Program—RALSP, International Center for Agricultural Research in the Dry Areas—ICARDA, Amman, Jordan

2 International Center for Agricultural Research in the Dry Areas—ICARDA, Directorate General of Agriculture and Livestock Research, Barka, Oman

3 Resilient Agricultural Livelihood Systems Research Program—RALSP, International Center for Agricultural Research in the Dry Areas—ICARD, Institut National de la Recherche Agronomique de Tunis—INRAT, Ariana, Tunisia

*Address all correspondence to: b.dhehibi@cgiar.org

References

[1] ICADRA. Steering Committee of ICARDA's Led Date Palm Project in Gulf Countries Meets in Doha. 2017. Available from: https://www.icarda.org/update/steering-committee-icardas-led-date-palm-project-gulf-countries-meets-doha

[2] FAO. Food and Agriculture Organization Online Dataset. 2018. Available from: http://www.fao.org/faostat/en/ [Accessed: April 2018]

[3] Frija A, Dhehibi B, Ben Salah M, Aw-Hasan A. Competitive advantage of GCC date palm sector in the international market: Market shares, revealed comparative advantages, and trade balance indexes. International Journal of Marketing Studies. 2017;**9**(6):1-13. Available from: http://www.ccsenet.org/journal/index.php/ijms

[4] Kohls RL, Uhl JN. Marketing of Agricultural Products. Upper Saddle River NJ: Prentice Hall; 1998. 560 p

[5] UN COMTRADE. United Nations Comtrade Database-International Trade Statistics-Import/Export Data. Online Dataset. 2018. Available from: https://comtrade.un.org/data/ [Accessed: April 2018]

[6] Mbaga MD. Date marketing. In: Manickavasagan A, Essa M, Sukumar E, editors. Dates: Production, Processing, Food, and Medicinal Values. Boca Raton, Florida, USA: CRC Press; 2012. pp. 155-172

Section 3

Resource Economics

Chapter 3

Effects of Water Scarcity on the Performances of the Agricultural Sector and Adaptation Strategies in Tunisia

Ali Chebil, Aymen Frija, Mariem Makhlouf, Chokri Thabet and Sihem Jebari

Abstract

The chapter aims to develop a regionally disaggregated agricultural supply model for Tunisia in order to investigate the potential effects of increasing water scarcity on the performances of the agricultural sector in the country, and the structural adaptation strategies needed to face such a challenge. A set of scenarios combining future water availability, water use efficiency, and increasing producer prices were simulated using the developed model. Results show that the agricultural sector in Tunisia, particularly the agricultural employment, would be negatively affected in case of decreasing irrigation water availability, and mostly affected regions would be the north east, central west, and southern areas. However, it is always possible to mitigate such effects through a combination of structural adjustments (changing land use in different regions), enhanced water use efficiency, and support of producer prices. The model also provides recommendations regarding specific crops that should be promoted in specific regions in order to maintain an agricultural sector with high added value in Tunisia.

Keywords: agricultural supply model, mathematical programming, regional level, water shortage, Tunisia

1. Introduction

The impacts of climate change (CC) will be channeled primarily through the water cycle [1], with consequences that could be large and uneven particularly on the agricultural sector. Ref. [1] also entails that some regions could see their growth rate decline by as much as 6% of GDP by 2050 as a result of water-related losses in agriculture, health, income, and property. For the Mediterranean countries, reduction of freshwater availability is predicted to attain more than 40% by the end of this century along the coastal areas [2]. The North African region is one of the regions which will be affected the most by CC, as anticipated by different climate models [3]. The region is already experiencing low rainfall characterized by its high variability, which is influencing agricultural production systems and changing their determinants. Climate model simulations are providing converging results concerning the decreasing trends of rainfall with 10–20% across North Africa [4], with average median decrease reaching 12% [5]. For Tunisia, this rainfall trend will result in a decline of water availability with up to 28% in 2030 [6, 7]. Ref. [1] also reports that water management policies can exacerbate the adverse growth impacts of CC, while good policies can go a long way toward neutralizing them. While CC is one of the major challenges facing humanity nowadays, adaptation frameworks to its, reversible and irreversible, impacts on the natural and human systems have emerged as an urgent need. It is expected to intensify risks related to natural resources availability, particularly in areas where water scarcity is already a concern [8]. In most countries, freshwater scarcity is increasing, forest fires are more frequent because of high temperature, drought is omnipresent and persistent, and desertification rates are growing [9]. Previous reports and analysis have described the Mediterranean region as a CC "hot spot" [10] including the Intergovernmental Panel on Climate Change (IPCC). Agriculture is a climate-sensitive sector subject not only to adverse impacts of CC on natural resources but also on social and economic contexts. Changes in precipitation and warming patterns are witnessed having occurred during the last century [11]. All year round widespread warming and reduction in rainfall are predicted by scientific literature for the twenty-first century [10]. Reduction in precipitation in addition to an increase in evapotranspiration would lead to water shortages particularly in regions where resources are already at a critical level and irrigated cropping areas are increasing. CC is thus contributing to narrowing the gap between water supply and demand [12] which entails more complexity on water resources management in agriculture [8]. CC is reshaping not only agriculture activity patterns but also driving human existence standards, which requires a restructuration of an institutional framework and a policy plan that could be able to mitigate and adapt to CC impacts. Therefore, exploring adaptive pathways [13] and climate policy is becoming a cross-scale central focus for decision and policy makers [14]. Ref. [15] demonstrated the role of regional, national, and global policies and institutions in highlighting adaptation options and tools [16] and that the development of CC adaptation as a policy field is considered as a relevant application context for the establishment of the agriculture policy [17]. In order to assess the implications of potential policy actions and to assist stakeholders in developing adequate measures to improve resilience to CC, [17] prevailed that cost-benefit analysis is a useful assessment tool; bio-economic models are more useful for an ex-ante evaluation of policy interventions by simulating agents' (farmers') behavior on the farm level. However, analyzing CC impacts on agriculture (economic, social, and environmental) requires an approach that is able to provide a detailed picture of the sector, its constituents, and the interactions within it. Agricultural models, can be built on micro-level; bio-economic models or macro-level; studies entailing the whole agricultural sector such as agricultural supply models. Agricultural supply model (ASM) provides a presentation of the agricultural sector by a sequence of behavioral equations whose objective is to maximize regional income subject to technological, environmental, and institutional constraints [17–19]. They treat a wide range of issues in agriculture; ASM has been used to predict and assess the impacts of Europe's Common Agricultural Policy (CAP) or to estimate economic value of water and land [20]. Assessing CC impacts on Tunisian agricultural sector is a propitious research field; hence, by means of an agricultural supply model, it is possible to assess the impact of water scarcity, engendered by CC, on the agricultural sector in the country.

In this chapter, we suggest to look to strategic structural adjustments needed in terms of land use and irrigation in Tunisia to deal with future water scarcity. Structural change in agriculture is defined as being the adjustment of the agricultural sector to the changing conditions of demand and supply [21]. This complex and dynamic process constitutes a reallocation of land use and farm specialization, as well as repositioning of the agricultural sector as compared to other sectors of the economy [21–23]. Within this general framework, the objective of this paper is to simulate the scope of future water scarcity scenarios on the agricultural sector of Tunisia and to provide recommendations on how to reduce its effects through a CC adaptive policy plan. For the following sections, we particularly refer to structural change as being the reallocation of land use and crop specialization among different regions in Tunisia, as well as upon rain-fed and irrigated conditions. A regionally disaggregated agricultural supply model for Tunisia (ASMOT) was developed and used to simulate the effects of declining irrigation water availability on the development of the agricultural sector in different regions of Tunisia. Implications in terms of regional agricultural value added as well as employment in both irrigated and rain-fed sectors were assessed under different water-related scenarios. To our knowledge, ASMOT is the first attempt of disaggregated sector modeling in Tunisia which we aim to further develop and validate in the coming years.

2. Agricultural sector in Tunisia

Agriculture is an important sector in Tunisia contributing to 8.7% of the national GDP and employing around 16.2% of the total employment in the country [24]. Major crops, in terms of cultivated area, are tree crops (especially olives and dates) followed by cereals. While tree crops are strategic for exports (Tunisia is among the top 5 world exporters of olive oil and dates), cereals remain very important for human and livestock domestic consumption. Tunisia is also characterized by low rainfall and limited renewable water resources. It is influenced by the arid and semiarid climate that covers more than three-fourths of its area [25]. The agricultural sector is also highly dependent on water resources since it consumes more than 75% of total water use in the country [26, 27]. Climate variability has major effects on agricultural production in Tunisia which results on highly variable yields along years. Other sectors might also be affected but certainly with much less extent. In fact, according to the Tunisian regulation, urban, industrial, and touristic sectors are prioritized in terms of water use during shortage periods. As an example of this fluctuation, total cereal production went from 2.9 million tons in 1996 to 0.5 million tons in 2002 and again to 2.9 million tons in 2003 [26]. This trend is observed for all cereal crops where the yield of durum wheat varies between 0.5 and 2 tons/ha, soft wheat yield ranges between 0.5 and 2.5 tons/ha, and barley yield is between 0.4 and 1.5 tons/ha. Not only yields are variable, but the cereal and fodder cropped areas are also depending stochastically on the climate conditions. For the expected "bad" years, farmers usually avoid planting cereals which engenders a decrease of both areas and yields. As strategic response to climate variability, the country has started since the early 1970s to expand its irrigated areas in order to ensure more reliable supply of agricultural commodities over the years [28]. This strategy partly succeeded in developing around 450,000 ha of irrigated areas representing around 8% of total agricultural area in the country. Although irrigated area share is low, it reflects the highest surface that can be irrigated by the available water resources, given the current levels of irrigation water use efficiency (IWUE). However, despite their low share in total agricultural land, irrigated areas in Tunisia are producing 35% of the agricultural value added, and they are contributing up to 20% of total agricultural exports and 27% of agricultural employment [26]. Around 48% of these irrigated areas are irrigated from groundwater sources, including both superficial and deep aquifers, allowing the irrigation of 48% of the total irrigated area [28]. Overall water resources in the country are estimated to be only around 4700 million m^3 [7] including 650 million m^3 of nonrenewable resources (13.8% of the total water resources). Surface water is estimated to 2700 million m^3. Another major problem of the agricultural sector in Tunisia is the small farms' size. In fact, average farm size in Tunisia in 2005 was only about 10.2 ha [27]. Total farm number is 516,000 farms, managing an area of 5.3 million ha. According to the same source, in 2005, 54% of these farms have a size lower than 5 ha, and 75% of farms have a size lower than 10 ha indicating the main structural problem facing the modernization of the agricultural sector and the irrigated areas. In this regard, the stabilization of agricultural yields and the decrease of the sector dependency to climate variations are thus necessary for enhancing food security and agricultural trade balance in Tunisia. Many solutions have been proposed including the improvement of farmers' skills, financing, mechanization, intensification, and the extension of the irrigated areas. A structural change, however, is a broader concept that permits the adjustment of agricultural sector not only upon market features but also a more sustainable management of natural resources, land and water, to reinforce resilience to climate variability and food insecurity. This paper actually aims to determine which national structural readjustments are relevant for a more efficient reallocation of resources using a country- and context-specific agricultural supply model and scenarios. The following sections explain in details the model structuring and also present and discuss the outcomes of the study.

3. Methodology and analysis

The ASMOT model is an agricultural supply model that is built based on primary and secondary data of farming inputs and outputs for different crops, regions, and systems (rain-fed and irrigated). ASMOT is the first regionally disaggregated ASM developed for Tunisia. The model includes 21 of the most strategic crops of the country (including the most important cereals, trees/fruits, and vegetables). It also includes a representation of 67% of the total agricultural areas of Tunisia (around 3.34 million ha) and 78% of the total irrigated areas (around 352,000 ha). The ASMOT model is built based on regional disaggregated data, including 24 governorates of Tunisia. These governorates have been aggregated into five regions (North West (NW), North East (NE), Center West (CW), Center East (CE), and South (SO)) based on bioclimatic homogeneity (**Figure 1**).

The model was calibrated through Positive Mathematical Programming (PMP) [29] and using official 2011 data about observed crop areas by region and system (irrigated/rain-fed) as recorded by the Ministry of Agriculture, Hydraulic Resources and Fisheries of Tunisia [30]. Regional irrigation water availability was also included into the model based on official secondary data about existing water reservoirs in the different regions of the country.

Regional agricultural value added are optimized by ASMOT and aggregated into a national domestic agricultural value added. Various types of biophysical and economic constraints are considered in parallel to this optimization process. These can be found in the next section presenting the main mathematical structure of the model. The model also considers crop evapotranspiration and their respective effect on yield gaps. The different crops and regions included in the ASMOT model are shown in **Table 1**.

3.1 Structure of the ASMOT model

The aggregated agricultural supply (Eq. (1)) of the model calculates the aggregated gross value of agricultural supply (AS) in Tunisia as the sum of regional agricultural gross production values (RAS). Eq. (1) can be read as follows:

$$AS_{c,s} = \sum_{r} RAS_{r,c,s} = \sum_{r} \{[P_c * (Y_{r,c,s} - \Delta Y_{r,c,s})] - [AC_{r,c,s} + WP_r]\} * X_{r,c,s} \quad (1)$$

where ASc,s is the total agricultural supply of different crops (c) and systems (s). Systems can either be rain-fed (rai) or irrigated (irr). RASr,c,s indicates the regional agricultural supply by region (r), crop (c), and system (s). Pc is the producer price of crop c; Y is the yield expressed by region, crop, and system; and

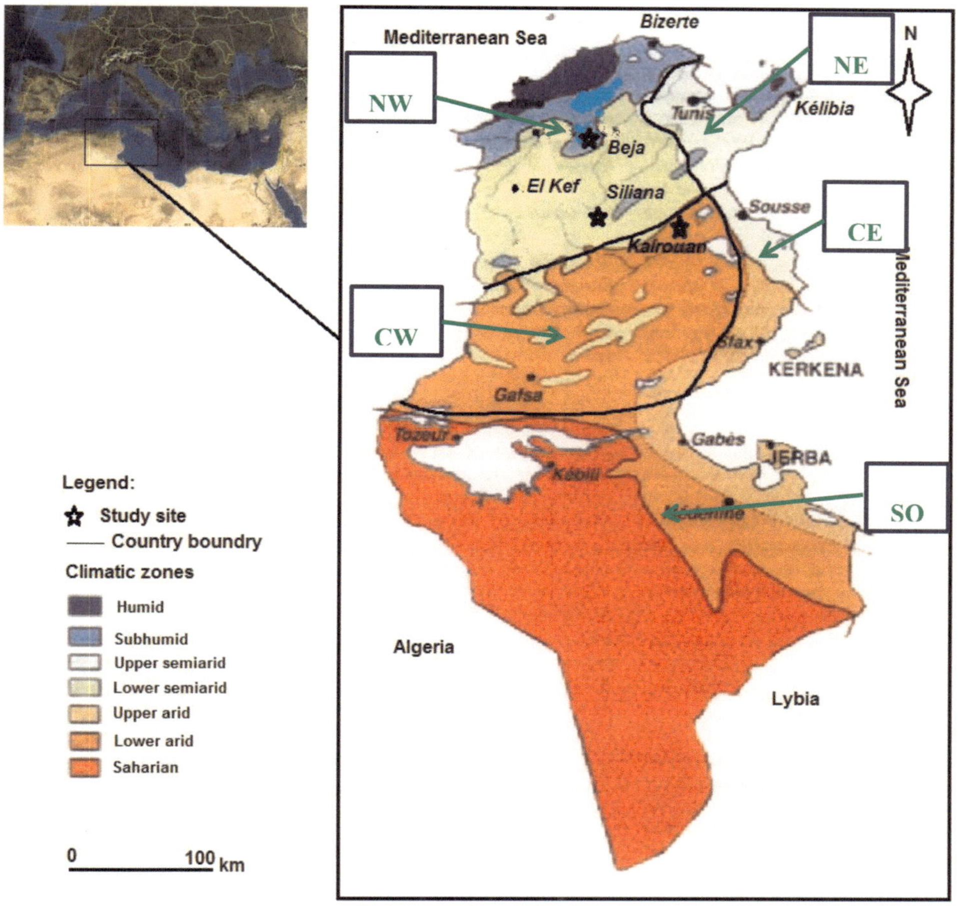

Figure 1.
Different bioclimatic regions in Tunisia.

Crops	Governorates and aggregated regions
Durum wheat, soft wheat, barley, olive, almond, palm date, citrus, grape, peach, apple, pear, grenade, tomato, potato, pepper, onion, garlic, artichoke, melon, watermelon, strawberry	North West (NW) (Bizerte, Beja, Seliana, Le Kef, Jendouba) North East (NE) (Nabeul, Ariana, Manouba, Ben Arous, Zaghouan) Center West (CW) (Sidi Bouzid, Kasserine, Kairouan, Gafsa) Center East (CE) (Sfax, Mahdia, Monastir, Sousse) South (SO) (Tozeur, Kebili, Tataouine, Médenine, Gabes)

Table 1.
Different crops and regions considered by the ASMOT model.

ΔY is the variation of yields which can be due to water stress (higher temperatures and evaporations). AC is the average cost of crop production excluding water costs. AC is expressed by region and system. WP is the irrigation water price in different regions. Finally, Xc,r,s is the positive variable of the total area for crop (c) under system (s) and in region (r). Observed Xc,r,s of the year 2011 was used for the calibration of Eq. (1). Once calibrated, X becomes variable and can be optimized under different scenarios. Yield variation ΔY is calculated as follows:

$$\Delta Y_{\mathrm{r,\,c,\,s}} = \mathrm{Y} * \mathrm{ky} * \left(1 - \frac{\mathrm{Eta}}{\mathrm{ETM}}\right) \tag{2}$$

where *ky* is the yield variation coefficient, which has a constant value for each crop, and Eta and ETM are, respectively, the real and maximal evapotranspiration:

$$\sum_{c,\,s} X_{r,\,c,\,s} \leq A_r \tag{3}$$

Constraint 3 is a land constraint, indicating that the total cultivated areas in each region should not, in the short term, exceed the currently observed agricultural areas (A_r):

$$\sum_{c,\,(s=irr)} X_{r,\,c,\,s} \leq IA_r \tag{4}$$

Constraint 4 indicates that the sum of crop irrigated areas in each region should not exceed the total irrigable areas (IA_r) available in that region:

$$\sum_{(c=trees),\,s} X_{r,\,c,\,s} \leq TA_r + (1 + \gamma_{c=trees}) \tag{5}$$

Constraint 5 bounds the annual tree area expansion to the observed annual growth rates of these areas in Tunisia during the last two decades which is about 5%. This constraint is also set at the regional level, where TAr is the current tree area in region r and γ is the annual growth rate of tree areas which is set to be equal to 5%:

$$\sum_{c,\,s} w_{r,\,c,\,s} * X_{r,\,c,\,s} \leq WA_r \tag{6}$$

Constraint 6 indicates that the sum of water requirement of all crops cultivated under different systems in a given region ($W_{r,c,s}$) should not exceed the water availability in that region (WA_r):

$$X_{r,\,c,\,s} \leq X^{o}_{r,\,c,\,s} * (1 + \varepsilon) \tag{7}$$

Finally, constraint 7 is a calibration constraint which was used in the first PMP step in order to estimate the cost function calibration coefficients ($\alpha_{r,c,s}$ and $\beta_{r,c,s}$). The average cost AC function is a nonlinear expression (Eq. (8)) estimated using two main calibration coefficients ($\alpha\alpha_{r,c,s}$ and $\beta_{r,c,s}$) which were calculated by solving Eq. (1) under the set of all considered constraints (3–7), including the calibration constraint [31, 29]. Coefficients $\alpha\alpha_{r,c,s}$ and $\beta_{r,c,s}$ were calculated using the dual values of constraint 7, and following the approach of [31, 32], where exogenous information about land rents was used for estimating the values of α and β. These PMP approaches have been widely validated and used for different sectors and other farm-type modeling and calibrations [33–39]:

$$AC_{r,c,s} = \alpha_{r,c,s} + \beta_{r,c,s} X_{r,c,s} \tag{8}$$

Eq. (8) was replaced by Eq. (1) which will generate a calibrated nonlinear objective function. To validate the calibrated model, we optimize Eq. (1) under all constraints while excluding the initial calibration in Eq. (7). If the resulting model will generate the same land allocation observed during the base year, then we can assume that our model is well validated and can be used for scenario simulations. ASMOT validation and calibration performances are presented in the result section.

3.2 Source of data

The data used for the ASMOT model was of different types and thus collected from various sources. Specific crop input and output levels for different regions and systems were collected through farmer questionnaires which were conducted for the season 2012–2013, in all regions of Tunisia in the framework of the Eau Virtuelle et Sécurité Alimentaire en Tunisie(EVSAT, funded by the IDRC) research project. Many focus groups with regional experts in crop production were conducted afterward in order to revise the average input and output values in respective regions and systems for all considered crops. Some coefficients of the model, such as the annual growth rates of tree crops, were calculated using FAO data [40]. Other secondary data regarding water availability, initial crop area distribution, irrigated areas, etc. were collected from official national datasets, especially available at the level of [30]. Water requirements in addition to evapotranspiration coefficients of different crops in different regions and systems were measured by the EVSAT research team through field experimentations.

3.3 Water scenarios

In relation to the overall objective of the chapter, our scenario development considers the current water scarcity situation faced by Tunisia, where water availability is expected to decrease by 28% at the end of the next decade [6]. Based on this, our first scenario suggests a cut of water availability by 25%, while second and third scenarios will consider improvements of IWUE and producers' prices as possible options to deal with this shortage and offer market incentives to enhance farmers' adaptation capacities. Only 69% of the total irrigated areas in Tunisia are fitted with water-saving technologies, thus leading to an average water use efficiency of about 55% at the national level [41]. This shows a wide scope to improve IWUE through appropriate investments in the farmer's skills and modernization of the irrigation networks. On the other side, it is well known that better integration of farmers along commodity value chains may offer enhanced producer prices [42], which can be considered as market incentives allowing farmers to enhance their technical investments and adaptation capacities [43, 44]. Based on these arguments, scenarios which were simulated using the ASMOT model can explicitly be read as follows:

- Scenario 1. Cutting total fresh water availability by 25%. This reduction is supposed to be the same across all regions of the country.

- Scenario 2. Cutting total fresh water availability by 25% and improving IWUE by 10%. The improvement of IWUE is interpreted in our modeling as a decrease of water volumes applied for different crops by 10%.

- Scenario 3. Cutting total fresh water availability by 25%, in addition to an increase of IWUE with 10% and higher producer prices offered to farmers. The suggested increase of producer prices are as follows: +5% for cereal prices and +10% for fruits and vegetable prices.

4. Results

After calibrating the model using real 2011 data and by estimating the calibration coefficients of the average cost function (Eq. (8)), the model was validated by running a status quo scenario and checking for consistency of the results compared to the observed values of land use. The result of this test showed that deviations of simulated land use variables ($X_{r,c,s}$) compared to the observed values of 2011 are all in the range of [−1%, +1%] (**Figure 2**) meaning that the model is performing well [37].

This validation test shows that ASMOT is performing well and can thus be used for scenario simulation. The next step is to reformulate and modify appropriate equations in the model in order to be able to simulate the scenarios presented in Section 2.3. Economic, social, and environmental outcomes of these scenarios are presented in the following sections.

4.1 Optimal land and water use under different scenarios

As discussed earlier, ASMOT optimizes the national agricultural value added and provides optimal regional land allocations for different crops and systems. These needed changes of land use in Tunisia allowing for optimal agricultural performances under a situation of water scarcity were purely calculated based on economic incentives corresponding to crop yields, costs, and incomes in the different regions and systems of Tunisia (**Table 2**). Results in **Table 2** show the overall trend of land use under different scenarios (SC1, SC2, and SC3).

Table 2 shows that trends of SC1 and SC2 are consistent but in most cases different from trends suggested under SC3. For the case of cereals, both SC1 and SC2 suggest important cuts of cereal areas in NW and CW and an increase of these areas in NE and SO regions. However, cereal areas are suggested to be reduced in all areas (except SO) under SC3. The same scenario 3 is also more favorable for expanding olives, almond, irrigated fruit trees, and vegetable areas in the different

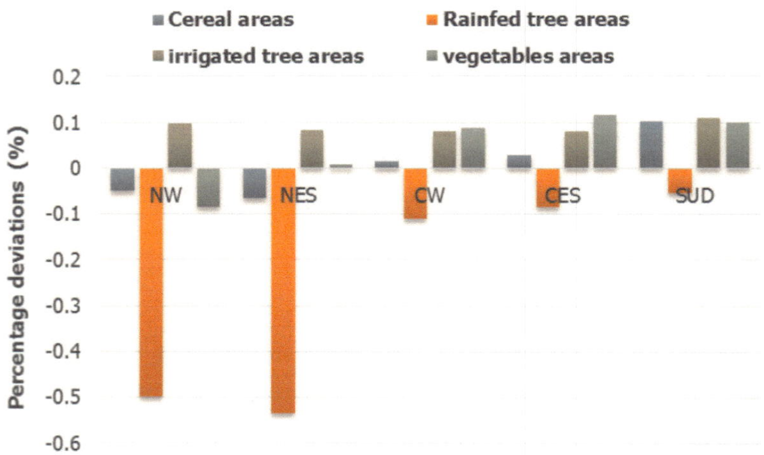

Figure 2.
Percentage deviation of simulated vs. observed crop areas in different regions included in the ASMOT model.

Type of crops	Regions	Percentage deviations compared to the status quo situation		
		SC1	SC2	SC3
Cereal crops	NW	−0.28	−0.06	−0.75
	NE	2.21	0.77	−1.39
	CW	−8.25	−3.37	−6.69
	CE	−0.01	0.03	−0.14
	SO	0.52	0.26	0.59
Olives and almond	NW	−0.8	−0.4	1.6
	NE	3.8	2.1	3.3
	CW	2.2	1.0	1.5
	CE	−0.1	0.0	−0.1
	SO	1.7	0.3	0.0
Irrigated fruit trees	NW	6.2	1.4	8.7
	NE	−8.0	−3.8	0.2
	CW	−5.2	−3.4	2.7
	CE	0.7	−0.6	3.4
	SO	−11.7	−2.3	−1.8
Vegetable crops	NW	8.0	2.9	4.8
	NE	−17.0	−7.3	−5.4
	CW	−4.1	−2.7	−0.6
	CE	7.5	2.7	7.5
	SO	4.6	2.2	8.1

Table 2.
Percentage change, compared to baseline situation, of the main crop areas under different scenarios (aggregated changes of rain-fed and irrigated systems).

aggregated regions. The highest area reductions recorded under SC1 and SC2 are these of cereals in CW; irrigated fruit trees in NE, CW, and SO; and vegetable crops in the NE. Under SC3, the highest area reductions were however recorded for cereals in CW and vegetable crops in NE.

4.2 Irrigation water demand under different scenarios

Total water use for irrigation under different scenarios in Tunisia was estimated based on optimal changes of land use as suggested in **Table 2** (see **Figure 3**). In the baseline scenario, around 2086.6 million m^3 of water is used for the total irrigated area considered in ASMOT (78% of the total irrigated areas, around 352.9 thousand ha) with an average use of 5912.1 m^3/ha (**Figure 3**). Under the first, second, and third scenario, total water consumption, respectively, decreases to 1876.5, 1818.1, and 1833 million m^3. By considering the new irrigated areas under each scenario, these decreases led to average water consumptions of 5949, 5349.7 , and 5385.8 m^3/ha. Total water saving under the second scenario is about 268.5 million m^3, which corresponds to around 13% of the total water use in the baseline situation. These numbers are showing that effective water management in the irrigated areas in Tunisia can mitigate the effect of water scarcity and even generate agricultural economic growth if accompanied by appropriate economic incentives.

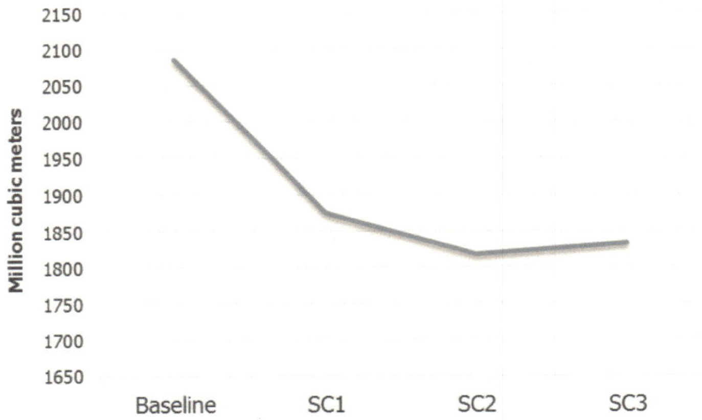

Figure 3.
Total water use for irrigation under different scenarios.

4.3 Impact on agricultural value added

ASMOT provides information about the total value added of its respective agricultural land area as the most aggregated results calculated based on optimization of these values at regional levels. This result can be calculated and presented for separate scenarios. For our particular case, the optimization process shows that Tunisia can overcome the problem of water scarcity (**Figure 4**) through specific structural changes of land use among crops, systems, and regions, as suggested in **Table 2**. **Figure 4** shows that agricultural value added in Tunisia will decrease with only 0.76 and 0.16%, respectively, under SC1 and SC2. However, these slight changes can only be possible if structural adaptations of the Tunisian agricultural sector, based on specific land use reallocations, are adopted as shown in **Table 2**.

Scenario 2 shows that with 10% increase of IWUE, the cut of water availability can be effectively mitigated, with an agricultural value added remaining almost equal to the status quo situation. If producer prices will further be supported (+5% for cereal crops and +10% for fruits and vegetable crops), the agricultural value added in Tunisia can even be 13% higher than the baseline situation, despite the sharp water cut considered. This higher value added of SC3 is not only due to the suggested price inflation but also to the restructuring of land use and the

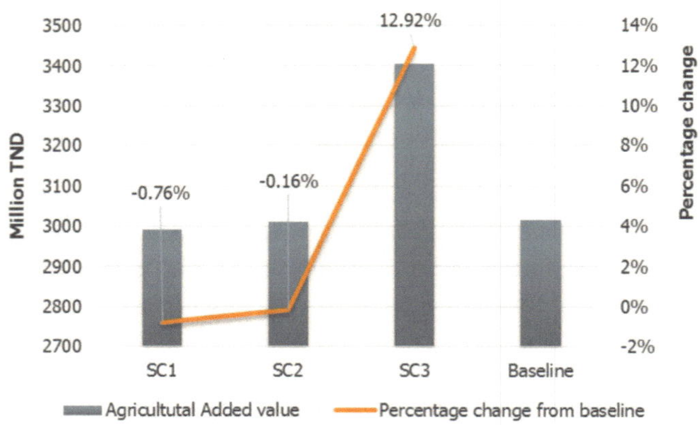

Figure 4.
Effect of water scarcity scenarios on the national agricultural value added.

decrease of total water use under this scenario. In fact, irrigated areas will decrease the most under SC3, and the average water use by hectare of irrigated land will also be 9.5% lower than SC1. Furthermore, the average price inflation considered under SC3 is only about 7.5%, with a maximum of 10% for vegetables and fruits. This price increase generated a higher and nonproportional increase of the value added (+13%), showing a relevant and positive and environmental effect of this price instrument.

Figure 5 shows a geographical distribution of changes in total agricultural value added among the considered regions, under different scenarios. It also shows the respective trends of these values among rain-fed and irrigated sectors. The figure shows that irrigated agriculture in Center West and North East of Tunisia is mostly affected by water scarcity. However, the contribution of the rain-fed agriculture in these two regions is also expected to grow which will partly overcome the negative effects of the decrease of irrigation value added.

4.4 Social effect of water scarcity scenarios

In this section we provide an overview of changes in labor demand under different scenarios compared to the baseline situation. **Figure 6** shows that despite the optimization of land use and agricultural value added, agricultural labor

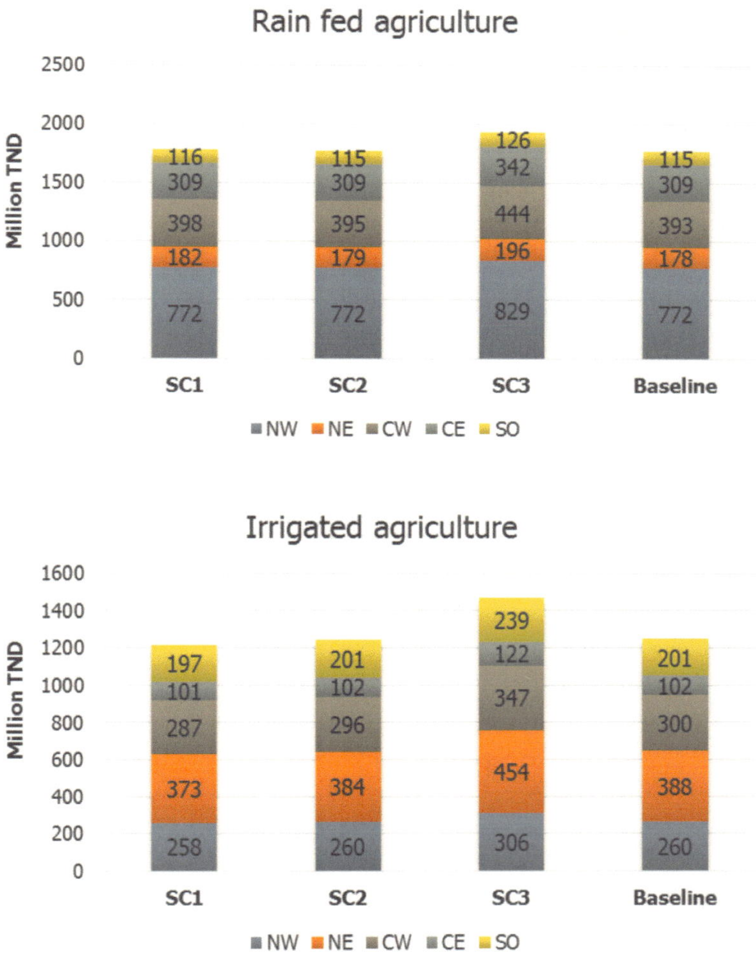

Figure 5.
Changes of regional agricultural value added under different scenarios in Tunisia (million TND).

demand will still be negatively affected under both SC1 and SC2, with respective decreases of 0.7 and 0.18% compared to the baseline situation. The same figure shows that this decrease of labor demand is exclusively recorded for irrigated areas and can reach–5.91% in these areas under the first scenario. The third scenario shows however that overall agricultural labor demand in Tunisia can increase with

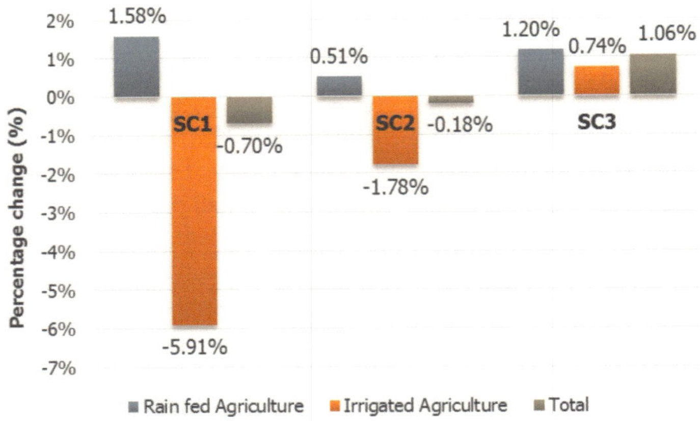

Figure 6.
Effect of different scenarios on regional agricultural labor demand (percentage changes compared to baseline).

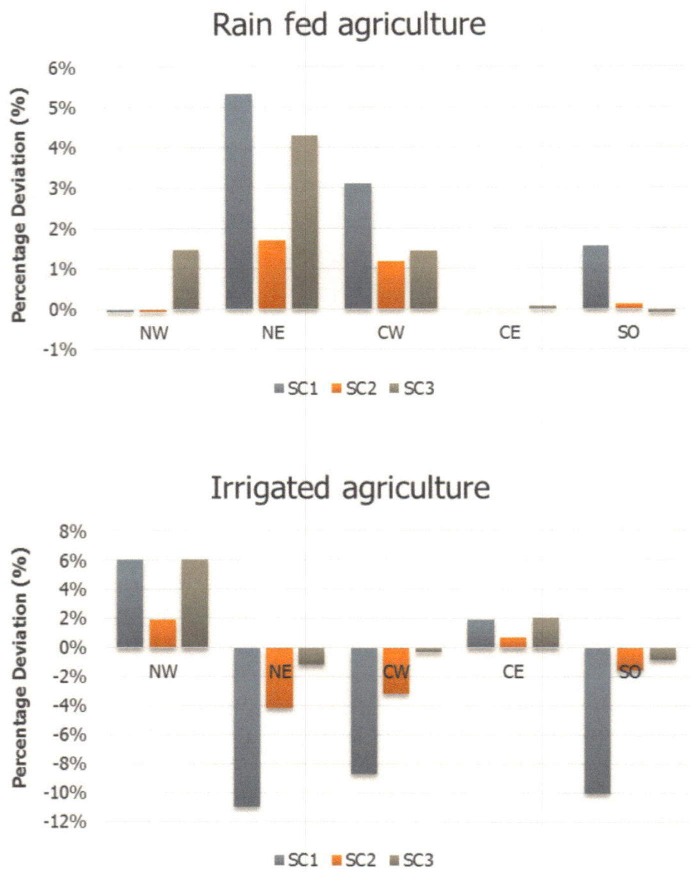

Figure 7.
Effect of different scenarios of regional agricultural employment under different scenarios (percentage change compared to baseline).

about 1.06% (around 8500 employment), despite the water scarcity situation. It is important to notice that, in opposite to SC1 and SC2, labor demand will increase under SC3 for the irrigation sector despite the decrease of the irrigated areas under this scenario (−3.6%).

Similar to the agricultural value added, labor demand in agriculture will disproportionally be affected along the different regions of Tunisia. **Figure** 7 captures most of these regional effects for both rain-fed and irrigated sectors. Despite the negative trend of labor demand in the irrigated sector, the restructuring of irrigated areas in the North West and Central East of Tunisia may generate slightly higher employment while at the same time maximize the value added of this sector. Furthermore, results show that labor demand in irrigated areas of South Tunisia will be decreasing even under the third optimistic scenario.

5. Discussions

The scope of enhanced IWUE was proven through our analysis to be highly effective in mitigating the effects of water scarcity in the different regions of Tunisia. Better IWUEs (SC2) are allowed for lower decrease of irrigated areas than the no IWUE scenario (SC1). In the NE region, these decreases were, respectively, −15 and −6% under SC1 and SC2. At the national level, irrigated areas decreased with −10.6 and −3.7%, respectively, under SC1 and SC2. This is showing a wide scope of IWUE to improve irrigation performances and sustain irrigation. However, IWUE can be defined at different scales including user/scheme and basin levels. Through our modeling framework, we only captured benefits of IWUE in terms of water saving. However, in addition to the benefits captured by our model in terms of water saving, physical efficiency at the user/scheme level will also be translated into increased water productivity (or economic efficiency) [45]. Mechanisms to reallocate saved water elsewhere in the water economy will further be necessary to enhance basin-level efficiency. On the other hand, only improvement of IWUE through better technology and management can generate real water savings [45]. Hence, in order to improve IWUE, some measures could be considered such as assisting farmers by providing enhanced knowledge about better irrigation scheduling of optimal amounts of applied water. Another measure would be related to better management of irrigation systems at the field and the landscape levels.

Without substantial improvement in the productivity of rain-fed agriculture, and despite a considerable expansion of cropped area, irrigated area would have to increase close to 500 million ha globally to meet the expected food demand, entailing a doubling of water use [46]. However, it is unlikely that suitable natural resources for such expansion might be available and the increase of agricultural productivity in both rain-fed and irrigated agriculture is necessary to meet such a global food demand. In Tunisia, our results show that rain-fed agriculture might be a good alternative for mitigating the effects of future water scarcity. In fact, value added of this sector was stable over the different scenarios, and it also showed a good potential for absorbing unemployment from the irrigated sector.

The overall effect of the water shortage scenarios on employment is negative, but this negative effect can widely be mitigated and improved if producer prices can be increased. Increased producer prices do not necessarily entail higher consumer prices but can simply be implemented through enhanced management, regulation, and control of agri-food value chains. This is in line with the suggestion that better integration of farmers along commodities value chains may offer enhanced and more equitable producer prices [42], which can in turn be considered as a type of market incentive for farmers and can be used to promote specific agricultural productions [43, 44].

6. Conclusions

In this chapter, we used an agricultural supply model to simulate the effect of water scarcity on agricultural production in Tunisia. We simulated three scenarios related to (i) cutting irrigation water availability, (ii) cutting irrigation water availability accompanied by relative improvement of irrigation water use efficiency, and (iii) scenario 2 in addition to enhanced producer prices for farmers. Results were overall showing that mitigating a shortage of irrigation water in Tunisia is possible through readjustment of irrigated and rain-fed areas and better allocation of crops among regions and systems (irrigated vs. rain-fed). Results also show that the best scenario which has a significant effect on agricultural value added is the third one. Under this scenario, agricultural employment in the overall agricultural sector can even increase. We strongly recommend that the "national agricultural map" already developed by the Tunisian government could be revised using further socioeconomic data and applied for an optimal allocation of crop areas across the country. We further recommend that more work should be done on better performing the structure and the functioning of the strategic agri-food value chain in Tunisia, allowing better marketing margins for farmers which will thus be translated into higher adaptation capacities of farmers to climate change and water scarcity.

Acknowledgements

This work was undertaken as part of, and funded by, the CGIAR Research Program on Policies, Institutions, and Markets (PIM) led by the IFPRI. Authors acknowledge the support of the Arab Fund for Economic and Social Development. The data used in this work was collected in the framework of the "Eau Virtuelle et Sécurité Alimentaire en Tunisie" project funded by the IRDC. The authors also acknowledge helpful funding from the European Union Horizon 2020 programme, under Faster project, grant agreement N°[810812].

Author details

Ali Chebil[1]*, Aymen Frija[2], Mariem Makhlouf[3], Chokri Thabet[4] and Sihem Jebari[1]

1 National Institute for Research in Rural Engineering, Water and Forestry (INRGREF), Ariana, Tunisia

2 Agricultural Center for Agricultural Research in the Dry areas, ICARDA, Ariana, Tunisia

3 National Agronomic Institute of Tunisia, Tunis, Tunisia

4 Department of Economics and Rural Development, Higher Agricultural School of Chot Mariem, Sousse, Tunisia

*Address all correspondence to: chebila@yahoo.es

References

[1] World Bank, High and Dry. Climate Change, Water, and the Economy. World Bank, Washington, DC. In: License: Creative Commons Attribution CC BY 3.0 IGO. 2016

[2] Abdulla F, Eshtawi T, Assaf H. Assessment of the impact of potential climate change on the water balance of a semi-arid watershed. Water Resources Management. 2009;**23**(10):2051-2068. DOI: 10.1007/s11269-008-9369-y

[3] Rochdane S, Bounoua L, Zhang P, Imhoff ML, Messouli M, Yacoubi-Khebiza M. Combining satellite data and models to assess vulnerability to climate change and its impact on food security in Morocco. Sustainability. 2014;**6**(4): 1729-1746. DOI: 10.3390/su6041729

[4] Schilling J, Freier KP, Hertig E, Scheffran J. Climate change, vulnerability and adaptation in North Africa with focus on Morocco. Agriculture. Ecosystems & Environment. 2012, 2009;**156**:12-26. DOI: 10.1016/j.agee.2012.04.0215N.I.C

[5] NIC-National Intelligence Council, North Africa. The Impact of Climate Change to 2030: Geopolitical Implications. Conference Report; 2009

[6] Paeth H, Born K, Girmes R, Podzun R, Jacob D. Regional climate change in tropical and Northern Africa due to greenhouse forcing and land use changes. Journal of Climate. 2009;**22**(1): 114-132. DOI: 10.1175/2008JCLI2390.1

[7] ITES – Tunisian Institute for Strategic Studies. Système Hydraulique de la Tunisie à l'horizon 2030. Tunis, Tunisia; 2014

[8] Iglesias A, Garrote L. Adaptation strategies for agricultural water management under climate change in Europe. Agricultural Water Management. 2015;**155**:113-124. DOI: 10.1016/j.agwat.2015.03.014

[9] Papadaskalopoulou C, Kasidoni M, Panaretou V, Moustakas K, Mesimeris T, Loizidou M. Review of the current EU framework on adaptation to climate change and assessment of the relative adaptation framework in Cyprus. Desalination and Water Treatment. 2016;**57**(5):2219-2231. DOI: 10.1080/ 19443994.2015.1107179

[10] Lionello P, Scarascia L. The relation between climate change in the Mediterranean region and global warming. Regional Environmental Change. 2018;**18**(5):1481-149. DOI: 10.1007/s10113-018-1290-1

[11] Lionello P, editor. The Climate of the Mediterranean Region: From the Past to the Future. London: Elsevier; 2012

[12] Wilhite DA, Sivakumar MVK, Pulwarty R. Managing drought risk in a changing climate: The role of national drought policy. Weather and Climate Extremes. 2004;**3**:4-13. DOI: 10.1016/j. wace.2014.01.002

[13] Haasnoot M, Kwakkel JH, Walker WE, ter Maat J. Dynamic adaptive policy pathways: A method for crafting robust decisions for a deeply uncertain world. Global Environmental Change. 2013;**23**(2):485-498. DOI: 10.1016/j. gloenvcha.2012.12.006

[14] Ford JD. Climate change research within regional environmental change. Regional Environmental Change. 2005; **15**(6):1035-1038. DOI: 10.1007/ s10113-015-0835-9

[15] Huq S, Reid H. Mainstreaming adaptation in development. Institute for Development Studies Bulletin. 2004;**35**: 15-21

[16] Rasul G, Sharma B. The nexus approach to water–energy–food security: An option for adaptation to

climate change. Climate Policy. 2016; **16**(6):682-702. DOI: 10.1080/ 14693062.2015.1029865

[17] Ponce R, Blanco M, Giupponi C. The economic impacts of climate change on the Chilean agricultural sector. A non-linear agricultural supply model. Chilean Journal of Agricultural Research. 2014;**74**(4):404-412. DOI: 10.4067/S0718-58392014000400005

[18] Howitt RE. PMP Based Production Models-Development and Integration. Copenhagen, Denmark: XI European Association of Agricultural Economists (EAAE); 2005. pp. 23-27

[19] Howitt RE, MacEwan D, Medellín-Azuara J, Lund JR. Economic Modeling of Agriculture and Water in California Using the Statewide Agricultural Production Model. Davis, CA: University of California; 2010

[20] Iglesias A, Garrote L, Martín-Carrasco F. Drought risk management in Mediterranean River basins. Integrated Environmental Assessment and Management. 2008;**5**(1):11-16. DOI: 10.1897/IEAM_2008-044.1

[21] Kirchweger S, Kantelhardt J. The dynamic effects of government-supported farm-investment activities on structural change in Austrian agriculture. Land Use Policy. 2005;**48**: 73-93. DOI: 10.1016/j. landusepol.2015.05.005

[22] Blandford D. Pressures for adjustment in the agricultural sectors of developed countries. In: Policy Reform and Adjustment in the Agricultural Sectors of Developed Countries. 2006. pp. 43-54

[23] Happe K, Hutchings NJ, Dalgaard T, Kellerman K. Modelling the interactions between regional farming structure, nitrogen losses and environmental regulation. Agricultural Systems. 2011;**104**(3):281-291. DOI: 10.1016/j. agsy.2010.09.008

[24] World Bank, Country Statistics. 2013. Assessed online on: http://data. worldbank.org/country/tunisia

[25] Lassaad L, Dhehibi B, Chebil A, Thabet C. National Agricultural Policy Report for Tunisia. 2005. Available from: http://medfrol.maich.gr/documentation/view/reports/wp1-napr/NAPR_Tunisia.pdf

[26] Dhehibi B, Aymen F, Aden AH. Performances, policies, challenges and opportunities of the tunisian agriculture sector from natural resources management perspective a SWOT analysis. American-Eurasian Journal Of Agricultural & Environmental Sciences. 2014;**14**:1351-1358

[27] MARH. Reports of Irrigated Perimeters in Tunisia. MARH; Tunisia. several years

[28] Frija A, Dhehibi B, Chebil A, Villholth KG. Performance evaluation of groundwater management instruments: The case of irrigation sector in Tunisia. Groundwater for Sustainable Development. 2015;**1**(1–2):23-32. DOI: 10.1016/j.gsd.2015.12.001

[29] Howitt RE. Positive mathematical programming. American Journal of Agricultural Economics. 1995;**77**(2):329. DOI: 10.2307/1243543

[30] ONAGRI. Official online database of agricultural productions, consumptions and markets. 2016. Accessible at: http://onagri.africadata. org/en

[31] Heckelei T, Britz W. Models based on positive mathematical programming: State of the art and further extensions. In: Arfini F, editor. Modelling Agricultural Policies: State of the Art and New Challenges. Proceedings of the 89th European seminar of the European Association of Agricultural Economics.

Parma, Italy: University of Parma; 2005. pp. 48-73

[32] Kanellopoulos A, Berentsen P, Heckelei T, Van Ittersum M, Lansink AO. Assessing the forecasting performance of a generic bio-economic farm model calibrated with two different PMP variants. Journal of Agricultural Economics. 2010;**61**(2): 274-294. DOI: 10.1111/ j.1477-9552.2010.00241.x

[33] Esteve P, Varela-Ortega C, Blanco-Gutiérrez I, Downing TE. A hydro-economic model for the assessment of climate change impacts and adaptation in irrigated agriculture. Ecological Economics. 2015;**120**:49-58. DOI: 10.1016/j.ecolecon.2015.09.017

[34] Gómez-Limón JA, Gutiérrez-Martín C, Riesgo L. Modeling at farm level: Positive multi-attribute utility programming. Omega. 2016;**65**:17-27. DOI: 10.1016/j.omega.2015.12.004

[35] Graveline N. Economic calibrated models for water allocation in agricultural production: A review. Environmental Modelling and Software. 2016;**81**:12-25. DOI: 10.1016/j. envsoft.2016.03.004

[36] Howitt RE, Medellín-Azuara J, MacEwan D, Lund JR. Calibrating disaggregate economic models of agricultural production and water management. Environmental Modelling and Software. 2012;**38**:244-258. DOI: 10.1016/j.envsoft.2012.06.013

[37] Medellín-Azuara J, Howitt RE, Harou JJ. Predicting farmer responses to water pricing, rationing and subsidies assuming profit maximizing investment in irrigation technology. Agricultural Water Management. 2012;**108**:73-82. DOI: 10.1016/j.agwat.2011.12.017

[38] Solazzo R, Pierangeli F. How does greening affect farm behaviour? Trade-off between commitments and sanctions in the Northern Italy. Agricultural Systems. 2016;**149**:88-98. DOI: 10.1016/ j.agsy.2016.07.013

[39] Torres MO, Howitt R, Rodrigues LN. Modeling the economic benefits and distributional impacts of supplemental irrigation. Water Resources and Economics. 2016;**14**:1-12. DOI: 10.1016/ j.wre.2016.03.001

[40] FAO dataset. Available online at: http://www.fao.org/faostat/en

[41] Blinda M. More Efficient Water Use in the Mediterranean. Working Paper n° 14, 2012. Plan Bleu, Valbonne. Plan Bleu Papers n° 14; 2012

[42] Trebbin A. Linking small farmers to modern retail through producer organizations – Experiences with producer companies in India. Food Policy. 2014;**45**:35-44. DOI: 10.1016/j. foodpol.2013.12.007

[43] Bahinipati CS, Venkatachalam L. What drives farmers to adopt farm-level adaptation practices to climate extremes: Empirical evidence from Odisha, India. International Journal of Disaster Risk Reduction. 2015;**14**: 347-356. DOI: 10.1016/j.ijdrr. 2015.08.010

[44] Mwinjaka O, Gupta J, Bresser T. Adaptation strategies of the poorest farmers in drought-prone Gujarat. Climate and Development. 2010;**2**(4): 346-363. DOI: 10.3763/cdev.2010.0058

[45] Turral H, Burke J, Faurès JM. Climate change, water, and food security. FAO report. Food and Agriculture Organization of the United Nations. Rome, Italy; 2011

[46] International Food Policy Research Institute (IFPRI). 2016 Global Food Policy Report. Washington, D.C.: International Food Policy Research Institute (IFPRI); 2016

Chapter 4

Local Techniques for Crop Conservation in Burkina Faso: Analysis of the Valorization Status and Perception of Tilgr-Baore Technology

Kala Brigitte Hema, Bienlo Annick Marina Paré and Marie-Thérèse Arcens Somé

Abstract

Agriculture is the main source of income for the poorest people and the pillar of food security in Burkina Faso. However, the producers face the problem of food insecurity each year, due to the effects of climate change and the difficulty of conserving their produce, because of inadequate storage infrastructure. These situations result in the loss of large quantities of products after harvest. Technological advances can provide solutions to this problem of postharvest losses and help many small producers to reduce poverty. Unfortunately, new product conservation techniques implemented in Burkina Faso are less known and poorly adopted. The objective of this chapter is to identify constraints to the valorization of new postharvest technologies of onion and potato and analyze factors influencing farmers' perception of these innovations. The results of the surveys conducted using the Tilgr-Baore technology show that many agricultural producers are analphabet and not yet aware of the existence of the new product conservation technology. The improvement of the level of education of farmers and the availability of information on innovations are needed to improve the perception of innovations and thus increase the probability of adoption of these innovations.

Keywords: agriculture, technological innovation, valorization, perception, Tilgr-Baore, Burkina Faso

1. Introduction

In Burkina Faso, agriculture is the main source of income for the poorest people and the pillar of the country's food security. It employs more than 80% of the active population and contributes to 30.3% of GDP [1]. Agriculture is dominated by small family farms, and production consists mainly of cereal crops, cash crops, legumes/tubers, and market garden crops. The market gardening crops concerns fruits and vegetables and is practiced especially in dry season. It accounts for 16.5% of agricultural production and generates about 400,000 jobs, 25% of which are held by women [2]. Market garden crop production is practiced in all regions of the country over an area estimated at 27,661 ha, with nearly 21 crops produced. Most of the vegetable production is for sale, with an overall marketing rate of over 90% [3].

The bulbous onion is the main market gardening culture both in terms of the area planted and the quantity produced. The total production of bulbous onion accounts for 32.4% of total vegetable production, and 41.4% of market garden areas are grown on bulbous onions [3]. The northern region is the main production area for onions. In this area, the potato also occupies an important place. Potato production in the northern region accounts for more than half of the national production [2].

However, market gardening production like most agricultural products in Burkina Faso is affected by the effects of climate change. The effects of climate change are mainly manifested by a decrease in rainfall, a deregulation of the rainy season, a greater irregularity of rainfall, and a frequency of drought [4]. Extreme phenomena such as floods and droughts are increasingly recurrent and affect farms. In fact, flooding affected 6.2% of the plots in operation during the 2016–2017 campaign, and 23.9% of the plots were affected by drought at the national level [5]. This rainfall variability affects the availability of water that is required for agricultural production.

In addition to these production risks, market garden production faces conservation constraints. Crop conservation is affected by insect rodent and mold attacks [6]. Crop storage infrastructures are insufficient and remain inadequate. These conservation difficulties are a major constraint for market gardening. Indeed, most market garden products are perishable, and poor conservation deteriorates their quality [7]. The lack of adequate conservation technology does not allow producers to store products in anticipation of periods when prices are more remunerative. Thus, the products are sold at harvest, and this does not ensure a balance between supply and demand throughout the year.

In order to ensure a good management of the market gardening product supply and to assure a good income for the actors of the sector, technologies allowing a better conservation of the products have been developed. However, these technologies are less known and poorly adopted. The objective of this chapter is to identify constraints to the valorization of these technologies and analyze factors influencing farmers' perception of innovations. Knowledge of the constraints to the valorization of new technologies and understanding factors affecting farmers' perception of these technologies are key in informing policies that aimed at spreading these technologies. The literature on adoption of innovation reveals that farmers' perception of the modern technology has significant influence on adoption decisions ([8] and [9]). Adoption of technologies by farmers may reflect rational decision-making based on farmers' perceptions of the appropriateness of the characteristics and the value of technology. However, the literature on socioeconomic factors determining such perception of technology is scarce.

The rest of the chapter is divided into six sections. Section 2 presents the configuration of the onion and potato market in Burkina Faso. Section 3 reviews methods and technologies for preserving agricultural products. Section 4 examines the technology valorization process in Burkina Faso. Section 5 analyzes the perception of farmers toward Tilgr-Baore technology. Section 6 concludes the chapter.

2. The configuration of the bulbous onion and potato market

The onion contains two products depending on the stage of the harvest. When it is produced for the purpose of harvesting only the leaves, we speak of leaf onion; on the other hand, if we wait for the maturity to have bulbs, we speak of bulbous

	Onion		Potato	
	2005	2008	2004	2008
Production (tons)	54,959	242,258	1376	6954
Area (hectare)	3681	11,449	73	262
Yield (tons/hectare)	15	21	18.85	26.54

Source: [10, 11]

Table 1.
Evolution of onion bulb and potato production.

onion. Bulb onion is grown in all 13 regions of Burkina Faso. Its production is seasonal and has increased considerably in recent years. Burkina's bulbous onion is marketed locally and in the subregion, especially in neighboring countries such as Cote d'Ivoire, Togo, Ghana, and Benin. In the rest of the chapter, we will use the term onion to describe the bulbous onion.

Like the onion, the potato production is realized mainly during the dry and cold seasons. The cultivation is necessarily irrigated, and the availability of a source of water is a sine qua non condition to its implantation. The production of potatoes is carried out by farmers grouped or not within producer organizations and is essentially intended to satisfy the national demand. **Table 1** shows the evolution of bulb onion and potato production in Burkina Faso.

The data in **Table 1** show that onion production increased by 341% between 2005 and 2008. This reflects the growing interest in the sector. It is an important source of income for rural people as more than 70% of production is devoted to marketing. The current situation of the local onion market is characterized by a single production cycle that runs from November to January with harvest periods that stop between March and April. This seasonality in onion production, combined with the weakness of onion conservation infrastructure, limits the availability of the product throughout the year. Fluctuations in the supply of onions during the year cause large price variations. **Figure 1** shows the evolution of the price of the kilogram of onion in CFA francs, during a year. Prices are highest from October to December declining thereafter and reaching their floor at the harvest period (in April). Thereafter they resume their cycle up until the end of the year.

Potato kilogram prices also fluctuate seasonally due to the low availability of the product throughout the year. Since the possibilities of conservation are very limited,

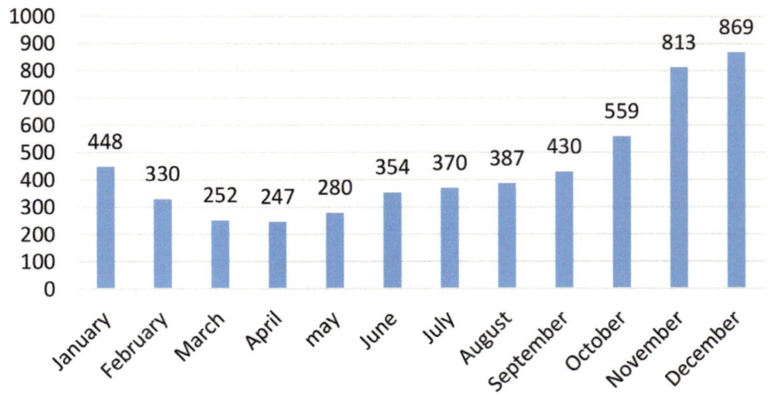

Figure 1.
Evolution of the price of the onion in CFA francs, during the year 2015. Source: [12].

the market is supplied irregularly, which has a significant impact on prices. The price of the kilogram of the potato drops sharply in times of plenty (150–200 FCFA/kg in February–March). A few months later, there is a rapid rise in price when the product becomes rare (between 500 and 700 FCFA/kg). At this point, it is the importation (from Mali and Europe) that intervenes to compensate for the insufficiency of local production [13].

The analysis of the onion and potato market pattern indicates that these two vegetable crops have identical production cycles. Almost all production is put on the market at harvest to avoid the losses that could result from poor conservation of products. Postharvest losses are highly dependent on the technologies used for storage and conservation. The example of the onion indicates that losses can reach up to 50% when production is not stored in any infrastructure [14]. The proportion of losses due to storage greatly reduces the profit margin of agricultural producers. The need to improve conservation techniques and the choice of appropriate structures is essential for the actors in the sector. However, what are the methods and technologies for the conservation of agricultural products in Burkina Faso?

3. Methods and technologies for the conservation of agricultural products

The Burkinabe government's efforts to promote agriculture that can stem food insecurity have earned the support of agricultural research to solve the problems that plague the sector. Thus, results of the research include not only the implementation of cultural methods but also crop conservation techniques to reduce postharvest losses and ensure availability of products. However, to cope with the problem of crop losses, the actors concerned use various techniques both traditional and modern to overcome the difficulties. Of a general nature, the drying of the harvests is a known practice but applies only according to the products. For cereals such as maize, millet, sorghum, and rice, farmers opt for exposure to the sun processing and the use of inert matter (ash and limestone) and repellent plants for their conservation [15].

Modern methods of conservation to reduce postharvest losses use products such as chemical insecticides (chlorpyrifos-methyl, pirimiphos-methyl, bioresmethrin, and deltamethrin) [16]. The transformation of certain products by drying from existing innovations such as solar drying equipment and storage warehouses makes it possible to preserve these products.

Among the conservation technologies, we find in the villages the granary that can store large quantities of crops. Designed from local materials such as animal dung straw, the attic is still used in rural areas to keep crops. For fruits and vegetables such as potato and onion, many producers use bags made from local materials such as straw. There is also the use of hangar and roofs and even houses to keep crops. Other technologies have been developed by local innovators. The innovators in Burkina Faso are mostly professionals (women and men) in the private sector working in workshops, small- and medium-sized industries, or small- and medium-sized enterprises, where they design, produce, and market new technologies. The main technological innovations for onion and potato conservation are the Klimax Z multipurpose fresh loft and the Tilgr-Baore storage box. The next section describes these two conservation innovations of agricultural products.

3.1 The Klimax Z multifunctional attic

The innovation Klimax Z still called the cool attic is a multipurpose solar cold room. **Figure 2** shows this innovation.

Figure 2.
The Klimax Z. Source: The promoter, 2018.

This storage infrastructure is designed and promoted by a young Burkinabe innovator. For its innovation, it benefits at the national level from the support of the Ministries of Scientific Research Energy, Youth Training and Professional Integration, and the Ministry of Agriculture.

The Klimax Z innovation has a capacity ranging from 6 to 60 m^3. It comes in two forms: the positive cold room (2–15°C) for the conservation of fruits vegetables and milk and the negative cold room (up to −10°C) for preservation of fish and other commodities. With a lifespan of at least 15 years, its cost of realization varies between 6 and 12 million CFA francs according to the minimal and large formats.

The Klimax Z can be installed in the markets for sellers of fruit and vegetables, meats, and fish, or it can be installed in the production sites (farm plots). This innovation can be exploited both in rural and urban areas. The exploitation of Klimax Z can be individual or collective. Associations or groups of actors such as producers and sellers in the markets can exploit it by proceeding by renting Klimax Z storage lockers. The shelf life of the potato lasts up to 9 months. In 2016 the Klimax Z was patented by the African Industrial Property Organization (OAPI) by PV Patent No. 17824 PV No. 1201600096 of January 22, 2016.

3.2 The box of conservation of the onion and the potato: the Tilgr-Baore

The Tilgr-Baore innovation is more specific to onion and potato conservation unlike Klimax Z, which takes into account a wider range of products. The glimpse of this innovation is presented in **Figure 3**.

The Tilgr-Baore box is a technology designed by a producer in the northern region of Burkina Faso. The designer of this technology is the head of an association ATPOY (Association Tickwende Yatenga Onion Producers) with the financial support from the Office of Cooperation Switzerland (BUCO) and GEDES-Burkina (the General Services a private consulting and engineering company). This innovation makes it possible to dramatically reduce the losses associated with traditional storage. It is a significant opportunity to overcome the problems associated with the conservation of onion and potato. The Tilgr-Baore storage box has a storage capacity of 7–10 tonnes and preserves the product for about 1 year. This innovation is made with local building materials (mainly straw) compared to cement storage warehouses. This makes the Tilgr-Baore box more accessible to onion and potato

Figure 3.
Presentation of Tilgr-Baore. Source: The promoter, 2018.

producers in rural areas. It keeps up to 98% of production. The cost of making this innovation amounts to 1 million CFA francs. Given this cost, the exploitation in association or in a group of both producers and sellers is recommended [14].

4. The valorization of new technologies for agricultural product conservation in Burkina Faso

This section presents the concept of valorization and analyzes the national environment through existing valorization structures and actions.

4.1 The concept of valorization

The valorization of the results of the research is a polysemic and multidimensional concept whose content varies according to the implied actors, their expectations, and their interests. [17] distinguishes six different categories from the concept of valorization in Burkina Faso: academic or professional, scientific, technological, economic, social, and political valorizations. Of course, these different categories are not mutually exclusive. Academic valorization involves the incorporation of recent research results into the training modules for updating the content of the courses offered to students or professionals engaged in the process of improving their professional knowledge and skills. Scientific valorization is mainly the activity of researchers. For these authors, the valorization of the results of research means the diffusion and the exchange of knowledge. The valorization is concerned in this case with the dissemination in the form of scientific publications in specialized journals, communications at conferences and seminars, posters, data sheets, photo, and video.

The social valorization of a research or an innovation aims to demonstrate the social utility of research or innovation in particular for the company to provide public funding. Economic valorization emphasizes the economic utility of innovation including its contribution to economic growth, increased productivity, and the creation of jobs and incomes for people. Regarding political valorization it aims to take into account innovations in the formulation execution, evaluation of public policies, and decision-making in general.

The national referential of the interventions of the government of Burkina Faso and its partners over the period 2016-2020, which is the National Economic and Social Development Plan (PNDES), gives importance to the valorization seeking to increase the number of research results and innovations valued in favor of the structural transformation of the economy. This is justified by the fact that most innovations are poorly known by potential users.

4.2 Analysis of the national environment

In Burkina Faso, scientific research and invention and innovation activities have generated many results that can contribute to the endogenous development of populations in many sectors of economic activity. Indeed the support provided to research and innovation since the 1990s by institutions such as the World Bank bilateral and multilateral cooperation agencies as well as international research institutions has enabled the country to have access to scientific and technological knowledge that can bring added value to the direct beneficiaries and to the State and thus reduce poverty in the country.

However, it is clear that many of these research and innovation results are less known to potential users, less used, and of little value. However, research activities are expensive, and the low value of research results is thus a poor allocation of resources as is the case in many developing countries [14]. In response to this, policy-makers are committed to enhancing the value of scientific research results, innovations, and inventions by institutionalizing the National Forum for Scientific Research and Technological Innovations (FRSIT) in 1995. The Ministry of Higher Education, Scientific Research, and Innovation (MESRSI) through the general direction of the National Agency for the Valorization of Research Results (DG ANVAR) organizes every 2 years the "FRSIT." This forum allows all actors (decision-makers, researchers, inventors and innovators, promoters, development partners, users, and the public) to become familiar with the research, its results, and the inventions and technological innovations realized in Burkina Faso but also outside the country. It is in this national environment favorable to the valorization of the results of research and innovation that the technologies Tilgr-Baore and Klimax Z were created by local innovators. What is the importance of intellectual property rights in the valorization of innovations?

4.3 The intellectual property rights

Intellectual property rights are mechanisms for the recognition and protection of intellectual materialization through invention and innovation which entitle their authors to use the fruit resulting from the valorization of their creation. Protection reassures innovators that products are not being pirated or spoofed. It puts innovators in trust with the use of their products with recognition because many innovators are reluctant to give information about the "processes" that are used in their innovation.

There are several systems of protection of an invention/innovation among which patents are the most known and most commonly used in Burkina Faso. The benefits of patent protection are multiple. In particular, the protection allows the work to be well scientifically documented, protects potential users, and provides substantial remuneration to the inventor or research institute concerned. Patents help to strengthen the country's brand image because the more the number of patents increases, the more the level and credibility of the research or innovation that produces them are enhanced [18].

Burkina Faso is a member of the African Intellectual Property Organization and the World Intellectual Property Organization. These institutions are responsible for the management of patents and other intellectual property rights. There is also a National Intellectual Property Office (DNPI) in the country which is the public institution charged with protecting the intellectual property rights of innovators and raising awareness and giving advice on the need for property ownership. However, the partnership of innovators with intellectual property organizations is weak in the agricultural sectors of Burkina Faso.

4.4 Valorization statements and constraint of valorization of Tilgr-Baore technology

The Tilgr-Baore technology was presented at the National Forum for Scientific Research and Technological Innovation (FRSIT) in 2016, where it was selected among the best innovations and awards. FRSIT has also included it on the national list of innovations. The steps for the patenting of Tilgr-Baore are in progress. The innovator with the support of FRSIT and DG ANVAR filed his patent application with the African Intellectual Property Organization. He holds from this organization for the moment just the acts which prove that he has introduced to OAPI his patent application for Tilgr-Baore.

Because of its importance for the conservation of agricultural products, the Tilgr-Baore is increasingly popularized by programs that work in the rural world. These projects include the Support Program for the Modernization of Agro-Pastoral Family Farms (PAMEFA program) whose objective is to promote modern family farming, the Agricultural Sector Support Project (PROFIL program) that involves installing 500 boxes of Tilgr-Baore types in four regions of the Burkina Faso, and an action research program of INSS, whose objective is to facilitate the access of Tilgr-Baore innovation to producers organized into associations and unions of associations.

Despite these promotional actions of Tilgr-Baore, there are constraints to the diffusion of innovation. The promoter of the innovation expresses itself in these terms: "There is the PROFIL program which has installed 500 boxes in 4 regions of Burkina Faso, but has not been completed. There was no follow-up or training for the operators of the box so that these beneficiaries could use the technology" (Ouedraogo Lassane, Ouagadougou, November 2018).

Considering the example of the PROFIL project, the developer estimates at 35% the completion rate of the construction of the 500 Tilgr-Baore box. Lack of training of beneficiaries remains the main obstacle to exploiting innovation. These remarks raise the difficulty of training beneficiaries on the various facets of technology. From the maintenance of the environment where Tilgr-Baore is installed, to the box itself and then to the stored products, a whole procedure of maintenance must be taught to the beneficiaries of the box for its judicious exploitation.

The Tilgr-Baore technology benefits from the support of various ministerial departments including the Department of Scientific Research and the Department of the Environment, Agriculture, Trade, and Crafts. This is an asset for promoting innovation, but it is difficult to bring together these different actors around a table to coordinate interventions. In this regard, the initiator of Tilgr-Baore said: "The fact that our technology brings together several ministerial departments, lead to difficulties to bring them together to work. This is a problem for us because we do not know how to bring all these authorities together to coordinate the actions of valorization of the technology" (Ouedraogo Lassane, Ouagadougou, November 2018). So, there is a need for a revitalization of the partnership between the ministerial departments, concerned with the promotion of innovation, and innovators.

Based on survey data from rural producers, the following section analyzes the perception of Tilgr-Baoré technology by rural people.

5. Perception of farmers toward Tilgr-Baore technology

Perception of innovation is very important in the adoption process. It indicates the user's view of a technology. Individual innovation decisions are based on the individual's perceptions of the innovation. According to [19], rejection, discontinuance, and re-invention frequently occur during the diffusion of an innovation, and such behavior may be rational and appropriate from the individual's point of view, if only the diffusion scholar could adequately understand the individual's perceptions of the innovation and of his or her own situation, problems, and needs. We are therefore analyzing the farmers' perception of Tilgr-Baore, in order to guide policies aimed at better dissemination of this innovation.

This section presents the model used to analyze the perception of Tilgr-Baore, the source of the data, the descriptive statistics of the sample, and the econometric results of the estimation.

5.1 Model specification and definition of model variables

The perception of Tilgr-Baore is analyzed using a probit model. The dependent variable is a binary variable. In most practical cases, one can choose indifferently between the models probit and logit, to analyze the perception of the innovation or adoption of innovation. We choose a probit model to analyze the perception of Tilgr-Baore technology.

5.1.1 Probit model

Consider Y^* the latent variable that is unobservable, whose value depends on a series of explanatory variables Xi. We have the following equation:

$$Y_i^* = \beta' X_i + \varepsilon_i \tag{1}$$

The dichotomous variable Y, observed, is linked to the latent variable Y^* by the following relation:

$$Y_i = 1 \quad if\, Y^* > 0 \text{ that is to say } \beta' X_i + \varepsilon_i > 0 \tag{2}$$

$= 0$ *ifnot.*

where the dependent variable Y_i is the farmers' perceptions of Tilgr-Baoré technology and X_i the sociodemographic and institutional characteristics that can influence the perception of innovation. β being the coefficients, ε_i is the error term. We assume the error term is distributed normally with mean zero and variance one. i = 1, 2......n. n is the number of observations.

If $\beta' X_i + \varepsilon_i > 0$, the farmer has a very good perception of the Tilgr-Baore, which tcould give him enough incentives to adopt the technology, and the dichotomous variable takes the value 1. The error term is due to the effects not considered.

5.1.2 Definition of model variables

The perception of the producer depends on several factors. The most conventionally analyzed are the socioeconomic and sociodemographic characteristics of the producer, as well as the institutional factors. The choice of the analysis variables of the model is based on empirical literature [20–22] as well as on the availability of data.

The dependent variable of the probit model is defined as binary variable, equal to 1 if the farmer has a very good perception of the Tilgr-Baore and 0 if not. The definitions and the measurements of the variables used in the estimation of perception were summarized in **Table 2**.

5.2 Source of data

The survey data were collected as part of the research program "Adoption and popularization of onion and potato conservation technology for food and nutrition

Variables		Expected sign
Dependent variable		
Farmers' perceptions	Level of appreciation of the technology. Equal to 1 if the producer has a very good appreciation of the technology and 0 if not	
Independent variables		
Age	Age is defined in terms of the number of years of birth. There are five age groups identified: Under 18 years, 18–34 years, 35–54 years, 55–70 years, and over 71 years	+/−
Sex	Sex is a binary variable equal to 0 if the producer is female and 1 if he is male	+/−
Level of education	The level of education is a binary variable that is equal to 0 if the producer has no level (cannot read or write) and 1 if the producer has acquired a level of education (by school or training adults)	+
Household size	Household size refers to the number of people living in the household. There are four household categories according to size. Small-sized households consist of less than 5 people. Then we have the household of 5–10 people, followed by household of 10–15 people, and finally the large household, with more than 15 people	+
Experience in farming	Experience refers to the number of years spent in agricultural activity. The different levels of experience are less than 5 years in agriculture, 5–10 years of experience, 10–15 years of experience, 15–20 years of experience, and more than 20 years of experience	+
Farm size	Area sown with onion and potatoes, in hectare	+
Information on the existence of new conservation techniques	Binary variable, equal to 1 if the farmer knows the existence of new technologies and 0 if not	+

Source: Authors, survey data, July 2018

Table 2.
Definitions of the variables used in the empirical model and the expected signs.

security in Burkina Faso: case of Tilgr-Baore." This program is part of an action research carried out by a multidisciplinary team composed of anthropologists, economists, sociologists, building specialists, and innovator of the Institute of Science of Societies (INSS) and the Research Institute of Applied Sciences and Technologies (IRSAT) of Burkina Faso.

This composition of the team has helped to collect quantitative and qualitative data required to cover the different lines of research. For the field phase, a sample of 300 producers was surveyed including 200 potato producers and 100 onion producers, using a questionnaire in the northern and north-central regions of the country. Data were collected in July 2018 at four sites: Yako, Tougou, Titao, and Zintenga. These sites are characterized by strong market garden productions. The specialization of the region in potato and onion production and the existence of selling opportunities in the country's capital and some neighboring countries have justified the choice of these areas. The producers' sample allowed us to carry out descriptive statistics and econometric estimates.

5.3 Descriptive statistics of the sample

The characteristics of the producers, the characteristics of the production, and the state of knowledge of Tilgr-Baore technology are presented in this section.

5.3.1 Characteristics of onion and potato growers

The sex and the level of education are the two mean characteristics of producers, presented here. **Figure 4** shows the distribution of the sample by sex.

The sample of producers is 57% female and 43% male. This means that women are more involved in market garden production. In the northern and north-central regions, market gardening is an income-generating activity for many women in rural areas, as the rural population is predominantly female. According to the continuous multi-sectorial survey [23], the rural population is 51.52% women versus 48.47% men. According to the level of education, **Figure 5** shows the percentage of producers according to the level of education obtained.

The analysis of the sample shows that the majority of onion and potato growers have no level of education. These producers represent a proportion of 53%. Of the producers, 25% are literate in French or in national languages in the non-formal education system and 18% have a primary level. Only 4% of producers reach a secondary level in their study. The lack of training of many producers can limit the use of technology.

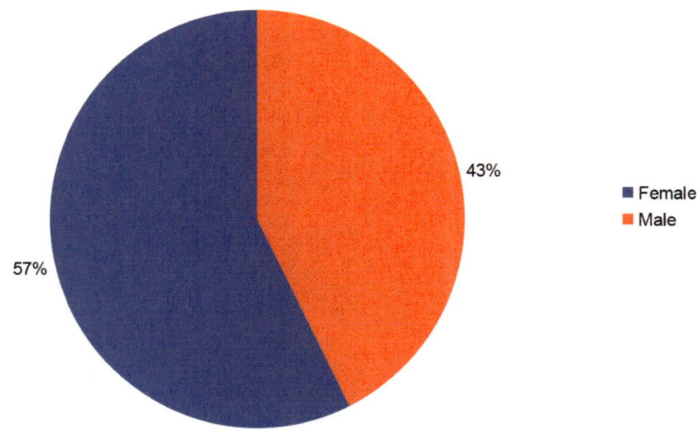

Figure 4.
Representation of producers by sex. Source: Authors, survey data, July 2018.

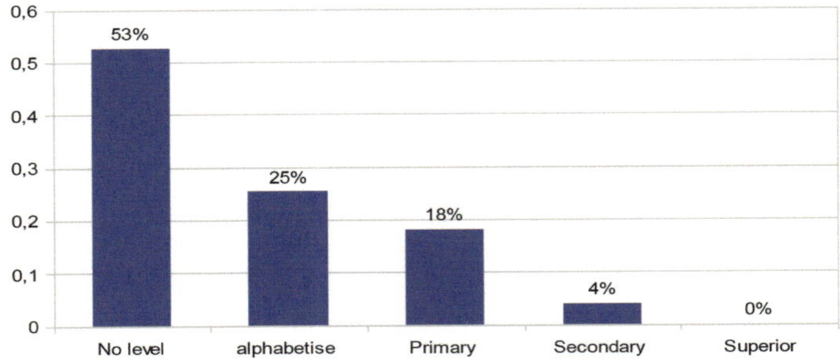

Figure 5.
The level of education of producers. Source: Authors, survey data, July 2018.

5.3.2. The characteristics of the production

This section highlights the perception of postharvest losses by farmers. **Figure 6** shows the percentage of producers facing production losses related to the conservation of the potato.

Potato producers record losses due to their conservation methods and technologies. Indeed 86% of producers experience losses of 21 kg or more related to the conservation of their potato production. The analyses show that the producers clearly perceive the losses related to their conservation technique. The extract from the interview with a producer from the study area proves that. This producer from the north-central region said: "There is no infrastructure to keep our production, we sell at harvest because it is difficult to keep. There is also the fact that those who produce the first, have an interest in selling it quickly. For the production from February to April, the market is very very weak, it is at this moment that we will gain to keep because it is at this period that we produce more. PAFASP[1] has tried to help some people build small sheds to keep, but it is very little compared to large production. So that we sell the production in the field because we can not preserve" (speech of Naaba Tingre producer of onions and potatoes at Zimtenga, July 2018).

5.3.3. State of knowledge of Tilgr-Baore technology

We present in this part the state of knowledge of the innovation, the source of the information about the technology, and the perception of the attributes of this innovation. **Figure 7** shows the state of knowledge of the Tilgr-Baore box in the sample.

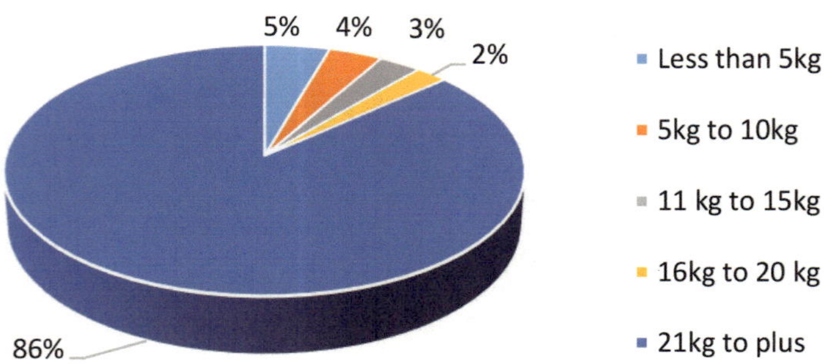

Figure 6.
The losses of the producers related to the conservation of the potato. Source: Authors, survey data, July 2018.

[1] PAFASP: Agro Sylvo Pastoral Support Program

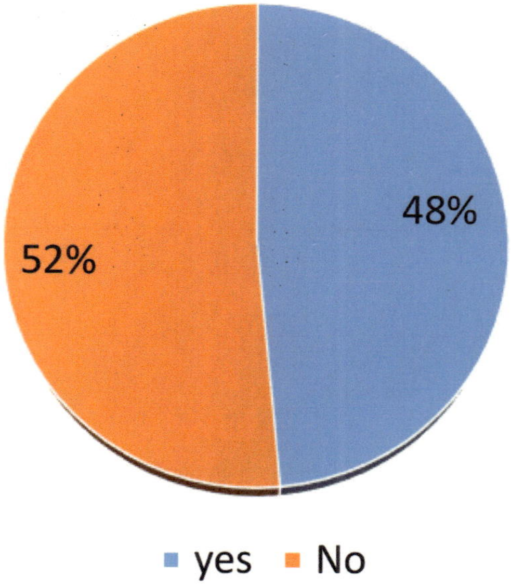

Figure 7.
Knowledge of the Tilgr-Baore box by the producers. Source: Authors, survey data, July 2018.

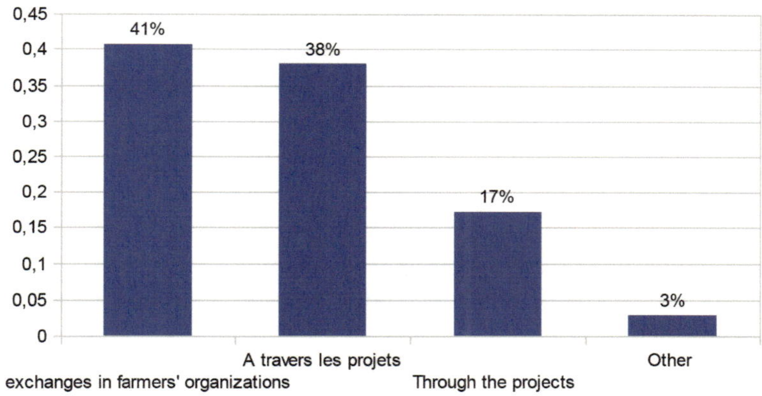

Figure 8.
Producer information channels on the existence of the Tilgr-Baore box. Source: Authors, survey data, July 2018.

More than half of the producers interviewed (51.5%) do not yet know the Tilgr-Baore conservation box. This ignorance of innovation does not allow producers to benefit from the advantages offered by this technology. What are the reasons that can explain this situation? Would it be a weak use of communication channels?

By focusing on the communication channels through which producers have learned about the box, **Figure 8** shows that the media played a weak role in disseminating information to producers. This could be explained by the high costs of disseminating information in the media or the low level of education of producers who are mostly illiterate.

Chat within farmer organizations is the most used communication channel for disseminating information on innovations. This mode of communication has allowed 41% of producers to be informed about the existence of the innovation. Only 17% of producers received information through the media channel (mainly radio). This result means that farmers' organizations play an important role in the dissemination of information because these exchange frameworks constitute elements of the social capital of producers. For a better promotion of the innovation,

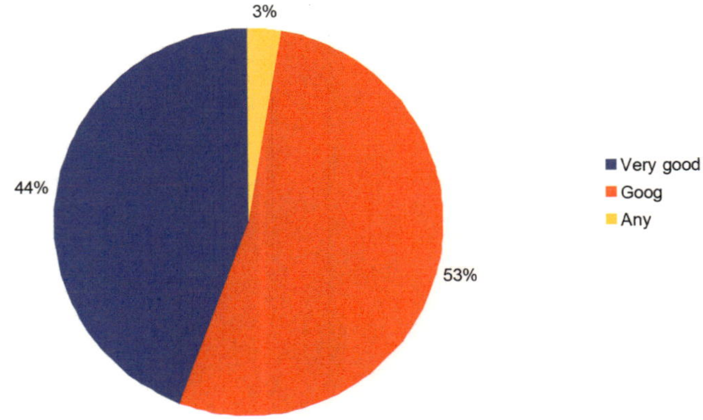

Figure 9.
The level of appreciation of Tilgr-Baore by the producers. Source: Authors, survey data, July 2018.

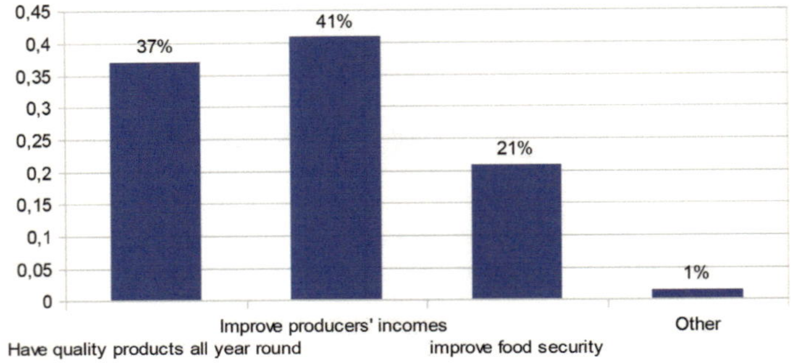

Figure 10.
The advantages of Tilgr-Baore according to the producers. Source: Authors, survey data, July 2018.

it is important that the diffusion projects are in relation with farmers' organizations and in contact with the producers. These actions would improve the level of appreciation of Tilgr-Baore by the producers. **Figure 9** gives an overview of the assessment of innovation by agricultural producers.

Among producers who are aware of the existence of the innovation, we note that 44% have a very good appreciation of the conservation box. The majority of producers surveyed (about 53%) gives a good appreciation of the box. However, to allow Tilgr-Baore technology to play an important role in the conservation and marketing of products, 55.9% of producers believe that its management should be done in groups or in association. In addition, a significant proportion of producers (41.4%) want the technology to be managed at the family level.

Producers evaluate the benefits of Tilgr-Baore in relation to increased income, improved product quality, and food security. **Figure 10** represents the perception of producers on the benefits of the innovation.

The result of producer's perception analysis on the benefits of the innovation show that 41% of producers find that the use of the conservation box improves incomes compared to 37% who believe that the technology maintain quality of products over a long period of time. Only 21% think that Tilgr-Baore improves food safety. This result implies that producers give more importance to the income-enhancing effect of the new conservation technologies.

The analysis of state of knowledge of Tilgr-Baore technology reveals that the technology is well appreciated and the benefits associated with its use are known. However, the proportion of producers who have this knowledge of technology is still low. This leads us to analyze the actions of valorization of the technology.

5.4 Econometric results of perception of Tilgr-Baore technology estimation

Table 3 presents the results of the probit model estimation of farmers' perception of Tilgr-Baore technology. The probit model parameters were estimated by the maximum likelihood method.

The likelihood ratio test (LR test) indicates that the estimated model is globally significant at the 1% threshold. Individual significance tests indicate that educational level, farm size, and information availability significantly influence the perception of Tilgr-Baore technology.

Educated farmers have a higher probability of perceiving the utility of Tilgr-Baore. When the producer is educated, the probability of having a very good perception of Tilgr-Baore increases by 12.71%. The level of education of the individual is an element of human capital. Instruction is described as an essential element in any development process. When a producer is educated, he can better appreciate the usefulness of new technologies, and he will be open to new ideas. The coefficient for size of cultivating land has a negative sign and shows that farmers with larger farm size were less likely to appreciate Tilgr-Baore technology. When the size of the farm increases by 1 hectare, the probability of having a very good perception of Tilgr-Baore decreases by 4.02%. This can be explained by the fact that small farms are the ones that face the most constraint of crop storage. The availability of information about the existence of new technologies has induced farmer more probability to perceive Tilgr-Baore technology. Access to information on the existence of new product storage technologies increases the probability of having a very good perception of Tilgr-Baore of 28.07%.

The positive sign associated with age, household size, and agricultural experience shows that these variables are positively related to Tilgr-Baore perception, but the effect is not statistically significant. These variables cannot explain the difference in perception of technology. Similarly, the sex variable has no significant effects on the perception of technology.

Perception of Tilgr-Baore	Marginal effects	Standard error
Age	0.0347137	0.0463629
Sex	−0.0344878	0.0636664
Level of education	0.1271664**	0.0589712
Household size	0.0429695	0.0315065
Agricultural experience	0.0359173	0.0300884
Farm size	−0.0402437**	0.019039
Information	0.2807886***	0.0664915
Number of observations	=	300
Likelihood ratio (LR) chi-square test: LR chi2(7)[1]	=	31.18
Prob > chi2[2]	=	0.0001

[1]*The likelihood chi-square test is a statistic test used for comparing the goodness of fit of two statistical models. The number in the parenthesis indicates the number of degrees of freedom. In this model, there are seven predictors, so there are seven degrees of freedom.*
[2]*Prob > chi² is the probability of obtaining the chi-square statistic, which is compared to a critical value 0.01, 0.05, or 0.1 to determine if the overall model is statistically significant. In this study, the model is statistically significant at the 1% threshold because Prob>chi² is less than 0.000.*
Mean the parameter is significant at 10%, ** the parameter is significant at 5%, and * the parameter is significant at 1%.*
Source: Authors, survey data, July 2018

Table 3.
Probit regression of farmers' perception of Tilgr-Baoré technology.

6. Conclusion

Onions and potatoes are market garden produce that are produced mainly during the dry season. The production of these two crops has increased sharply in recent years as it represents a major source of income for rural populations in Burkina Faso who are mostly poor. The current situation of the local market for these products characterized by a single production cycle makes it difficult to offer products throughout the year. A large part of the production is put on the market of the harvest to avoid the losses postharvest due to the lack of adapted storage infrastructures. To improve farmers' profit margins and maintain product quality throughout the year, local innovators have created new conservation technologies. These new technologies need to be valorized because they are not sufficiently known by the users who are the producers and the sellers. The valorization of technological innovations must aim for a structural transformation of the economy. The analysis of the national environment shows in a general way an environment favorable to the valorization of the innovations. The results of the surveys show that many agricultural producers are not yet aware of the technological innovations of onion and potato conservation. These producers are weakly alphabetized. The lack of training of many producers and the high cost of innovations can limit the use of innovative technologies. Econometric estimation of the determinants of the farmers' perception of Tilgr-Baore technology indicates that educational level and information availability have a positive and significant influence on the perception of Tilgr-Baore, while farm size negatively influences the probability of a very good perception of technology.

For a better dissemination of new technologies for onion and potato conservation, public policies must revitalize the partnership between ministerial departments involved in the promotion of innovations and innovators. In addition, public policies should encourage the promotion of innovation protection and speed up the process of obtaining patents to ensure the transfer of technological innovations. The adoption of innovations requires an improved perception of innovation. To do this, it is necessary to improve the level of education of farmers and make available information on innovations. Farmers' organizations are the main channel for transmitting information on the existence of innovations. It is an element of the social capital of the farmer that needs to be strengthened. Our results also imply that policies to promote new technologies adoption (mainly the Tilgr-Baore) should target small farms as a priority.

Author details

Kala Brigitte Hema, Bienlo Annick Marina Paré* and Marie-Thérèse Arcens Somé
INSS/CNRST, Ouagadougou, Burkina Faso

*Address all correspondence to: kaboreparemarina@gmail.com

References

[1] Plan National de Développement Economique et Social (PNDES); 2016-2020, Burkina Faso. 88p

[2] Ministère de l'Agriculture de l'Hydraulique et des Ressources Halieutiques (MAHRH) 2007. «Analyse de la filière maraichage au Burkina» / DSA/DGPSA/MAHRH/FAO 127p

[3] Direction de la Prospective et des Statistiques Agricoles et Alimentaires (DPSAA); 2011. Rapport général du module maraichage. 318p

[4] Ouédraogo M, Dembélé Y, Somé L. Perceptions et stratégies d'adaptation aux changements des précipitations: cas des paysans du Burkina Faso. Sécheresse. 2010;**21**(2):87-96

[5] Ministère de l'Agriculture et des Aménagements Hydrauliques (MAAH) 2017. Rapport général des résultats définitifs de la campagne agricole 2016/2017 et des perspectives de la situation alimentaire et nutritionnelle. DGEES ? 95p

[6] Waongo A, Yamkoulga M, Dabire-Binso CL, BA MN, et Sanon A. Conservation post-récolte des céréales en zone sud-soudanienne du Burkina Faso: Perception paysanne et évaluation des stocks. International Journal of Biological and Chemical Sciences. 2013;**7**(3):1157-1167. Available from: http://indexmedicus.afro.who.int

[7] Ministère de l'Agriculture de l'Hydraulique et des Ressources Halieutiques (MAHRH). Analyse des résultats de l'enquête maraichère campagne 1996 à 1997. Direction des Statistiques Agricole Mars; 2003. 85p

[8] Negatu W, Parikhb A. The impact of perception and other factors on the adoption of agricultural technology in the Moret and Jiru Woreda (district) of Ethiopia. Agricultural Economics. 1999;**21**:205-216

[9] Adesina AA, Baidu-Forson J. "Farmers' perceptions and adoption of new agricultural technology: Evidence from analysis in Burkina Faso and Guinea", West Africa. Agricultural Economics. 1995;**13**:1-9

[10] Guissou R, Cissé K, Pouya T. Analyse des incitations et pénalisations pour l'oignon au Burkina Faso. Série notes techniques. Rome: SPAAA, FAO; 2012. 41p

[11] Ministère de l'Agriculture et de la Sécurité Alimentaire, (MASA) 2014. Annuaire des statistiques agricoles 2012. Burkina Faso. 227p

[12] Institut National de la Statistique et de la Démographie, (INSD). Annuaire Statistique 2016. Burkina Faso: Ministère de l'Economie, des Finances et du développement; 2017. 379p

[13] Comite permanent Inter état de Lutte Contre la Sécheresse dans le Sahel (CILSS). Étude diagnostique de la filière pomme de terre dans trois pays de l'Afrique de l'ouest, Cas du Burkina Faso; 2003. 45p

[14] Parvex F, Somé G, Tossounon A. Une innovation qui libère: Capitalisation de l'expérience du grenier de conservation Tilgr-baore et de ses effets sur la filière de l'oignon et sur l'égalité des genres. SEREC; 2017. 36p

[15] Sankara F, Sanou AG, Waongo A, Somda M, Toé P, Somda I. Pratique paysanne post récolte du maïs dans la région des Hauts-Bassins du Burkina Faso. Journal of Animal & Plant Sciences. 2017;**33**:5274-5288. Publication date 3/07/2017, http://www.m.elewa.org/JAPS; ISSN 2071-7024

[16] Somda KBM. Problématique de la conservation post récolte du maïs dans les Hauts Bassins. Mémoire d'ingénieur de conception en vulgarisation agricole. Université polytechnique de Bobo Dioulasso BF; 2016; 65p

[17] Butare I. La valorisation des résultats de recherche et de l'innovation comme facteur de développement en Afrique: l'exemple du Burkina Faso. Colloques sur les innovations au Burkina; 2010. 14p

[18] Centre d'Analyse des Politiques Economiques et Sociales (CAPES). Résultats de la recherche, de l'invention et de l'innovation technologique au Burkina Faso: Etat des lieux de l'exploitation pour un développement durable. Burkina Faso; 2006. 56p

[19] Rogers EM. Diffusion of Innovations. The Free Press. A Division of Macmillan Publishing Co., Inc; 1983. 439p

[20] Patidar S, Patidar H. A study of perception of farmers towards organic farming. International Journal of Application or Innovation in Engineering & Management (IJAIEM). 2015;**4**(3):269-277

[21] Ochenje IM, Ritho CN, Guthiga PM, Mbatia OLE. Assessment of Farmers' Perception to the Effects of Climate Change on Water Resources at Farm Level: The Case of Kakamega County, Kenya. African Association of Agricultural Economists, Addis Ababa, Ethiopia; 2016. 16p

[22] Karunathilaka SDDD, Thayaparan A. Determinants of farmers' perceptions towards the adoption of new farming techniques in paddy production in Sri Lanka. Journal of Economics and Sustainable Development. 2016;**37**:-45

[23] Institut National de la Statistique et de la Démographie, (INSD). Enquêtes multisectorielles continues (ECM) 2014: Caractéristiques sociodémographiques de la population; 2015. 57p

Section 4

Environmental Issues

Chapter 5

Value Chain-Induced Constraints Limiting Scale of Conservation Agriculture in South Africa

Wolfgang Johann von Loeper, Scott Drimie and James Blignaut

Abstract

The potential of scaling conservation agriculture (CA), for long-term food security, remains under-investigated within the context of agricultural food value chains in South Africa. To scale the use of CA an understanding of the current agricultural value chains, their functioning, regulatory framework and constraints, is essential and this raises a key question: What are the main shortfalls and deterrents in agricultural value chains and why might CA be faced with challenges to feed into these existing structures, through which it could, the hopes are, create a more inclusive and sustainable farming system for long-term food security? The empirical data from an ethnographic qualitative participant research showed that interviewed value chain participants (VCP) are limited in acting on account of their economic constraints. None of them had products that supported CA, while financial institutions argued that such products would not be necessary, as any risk mitigating farming system would, in any event, result in financial benefits to the farmer.

Keywords: agricultural value chains, sustainable agriculture, agricultural economics, agricultural finance, farm ecology & policy

1. Introduction

Biodiversity is the planet's greatest asset [1]. Anthropocene-induced species loss is estimated at up to 10,000 times the rate of natural extinction, in which Hui et al. [1] argue agriculture, next to overfishing, industrialisation and urbanisation, plays a considerable role. Humans rely heavily on ecosystem services, which include cleaning air and water, stabilising weather, maintaining soil fertility, dissipating waste, controlling pests, pollinating crops, generating power and discovering new antibodies, and providing food, timber, cloth, medicine, minerals and industrial materials such as coal, oil, gas, rubber, plastics, and chemicals [1]. Humans have never contributed to such flows, but have always made use of them, today at a rate, where such ecosystem services are less likely to be available indefinitely.

Planetary boundaries, a concept developed by Rockström et al. [2], which identifies safe operating spaces within earth systems, such as climate stability, fresh water, land system change, ocean acidification, phosphate and nitrogen biochemical flows and biosphere integrity, are integral parts of the ecosystem services and represent our planet's limits in supplying such services within the principles of our planet's carrying capacity. To sustain humanity, we need to manage its biosphere within that carrying capacity, to maintain such services, and avoid regime shifts, mass extinction or repeating boom-bust patterns of earlier civilisations which were unable to manage their natural resources and regional carrying capacities [1].

1.1 Problem statement

Of Rockström's et al. [2] and Steffen's et al. [3] eight planetary boundaries, agriculture is by far the biggest contributor to defined limits of five of the boundaries; fresh water use, climate change, change in nitrogen and phosphate bio-chemical flows, land-use change as well as biodiversity loss. Agriculture also contributes up to 30% of CO_2 emissions to climate change [4, 5] and is, due to feedback loops from nitrogen and phosphate bio-chemical flows and deforestation, also a great contributor to biodiversity loss [2, 3].

Nelson et al. [6] suggest that due to climate change, global agricultural output is likely to decline between 10 and 15% in the next 60–70 years and even up to 50% in drier regions of Africa. Compared to the rest of the continent, arguably, much of South African (SA) agricultural land is located within such dry regions. With predicted changes, SA might need to consider whether its conventional farming (CvF) systems are appropriate going forward, while on the other hand evidence shows that alternatives, more sustainable farming systems such as CA, are comparably more climate resilient [7–9]. Arguments that farmers should adapt to such production systems in order to mitigate an output reducing impact due to climate change are weighing in more and more.

Low tillage, a form of CA regularly practiced in KwaZulu-Natal (~60%) and the Western Cape (>70%), indicates that in two provinces good headway has been made in favour of CA; yet finds little to no adoption in other provinces [10]. CA is based on three principles, no-till, crop rotation, and cover crops (residue retention) to increase both soil organic matter, aggregate stability and water holding capacity, while reducing soil bulk density, erosion, carbon emission and exposure to drought and ultimately increased yield [11]. With rain-fed crops in dry climates, CA can significantly increases productivity [12]. Pittelkow et al. [12] also argue that this indicates that CA will play an important role in mitigating the impacts of climate change. Therefore, CA is one of many farming practices farmers can adopt to farm with less environmental impact, while preparing for climate change.

Midgley et al. [13] argue that while South Africa's National Development Plan has identified agriculture as a primary economic objective, although not explicit, it is biased towards large scale, commercial and CvF practices, such as tillage and monoculture. South Africa's Integrated Growth and Development Plan [14], as well as the Agricultural Policy Action Plan [15], on the other hand, promote equitable growth and sustainable use of resources.

Food security is defined as having access to food of nutritional value at all times [16]. In this article we argued that CvF in a world of climate change poses a risk to food security, while a focus on more sustainable farming practices such as CA uses less water, requires less nitrogen and phosphate, sequesters CO_2 and diversifies the ecosystems of farmland, with the ability to decrease soil erosion, increase soil life and fertility and other ecosystem services to the benefits of a farmer's long-term profitability [11, 13, 17, 18]. Its uptake, however, remains low in SA. We argue that CA has an important role to play in a transition and show why, from evidence of our research, CA does not find support from SA food value chains.

This leads to questions such as: why CA adoption rate remains low; what role agricultural VCPs can play to promote CA; and what institutions, policies, and VCPs are responsible for hindrances to adoption? What limitations do VCPs themselves experience in potentially supporting CA, and how do feedback loops in existing business models of VCP block a transition to CA?

1.2 Research approach and design

Our study was undertaken as an ethnographic based research exploring business cultures and morals using qualitative semi-structured interviews. The questions for the research participants (VCPs) were not directed at any commodity in particular; however, because we also questioned silo owners and millers of maize, answers of some VCP often hinged around maize, also a main crop type in South Africa [19]. The choice of businesses interviewed was based on their involvement in the food value chain and their general size and importance they played and impact they had in their respective industries. Because of the sensitivity of the topic the interviewer needed to let go of any presumptions and assumed a less critical and more supportive attitude to attain more unbiased responses from the participant.

The interviews were then transcribed to attain primary qualitative data. For the coding and categorising, we used grounded theory as an inductive systematic methodology typically used in social sciences to analyse qualitative data and give it conceptual structure through categorisation of general themes emerging from the data [20–23].

Preceding the analysis and results of the research data, we reviewed literature to assess existing knowledge around the challenges facing existing economic and ecological farming systems and relate it back to CA.

2. Literature review

Conventional agricultural systems, particularly practiced in the developed world, produce vast amounts of food, yet they come at a significant cost to the environment. While the situation is complex, the details are often not acknowledged; in the following we outline high level important aspects that challenge the long term economic, social, and ecological sustainability of CvF and then show what alternatives exist that could replace CvF practices.

2.1 Problems of conventional agricultural systems

Covering 1/3rd of the planet's surface [24] agriculture has resulted in disturbed ecosystems [25–27], land degradation [28], loss of biodiversity [29–31], leaching fertiliser, nitrification of groundwater, eutrophication of above groundwater ecosystems, coastal dead zones [26, 32], small organism mortality [33, 34], and biological resistance build-up against agrochemicals [35–37]. Modern industrialised agriculture and overgrazing are blamed for destroying a third of the planet's topsoil within 40 years, adding 10 million hectares every year to the toll of soil erosion [38] which is 100 times faster than naturally occurs [39].

We deploy 2½ million tons of pesticides and fungicides annually and nevertheless lose 40% of crops globally to pests, diseases and weeds [40], while its use is also responsible for over 40,000 human deaths and 3–5 million cases of pesticide poisoning every year [41]. At no time in history has agriculture had such a high impact on the environment than in the last 100 years [25, 42–44].

2.2 Resource hungry agriculture's impact on planetary boundaries

Agriculture globally occupies 13 times more land than any other Anthropocene land use [24] and is arguably the biggest contributor to biodiversity loss and altogether the greatest human impact on our "planetary boundaries" [2]. Seventy percent of all freshwater globally is used for agriculture [41], while in SA it is estimated to be 63% [45] with no surplus for future development [46]. Agriculture is energy-hungry, emitting up to 25% of global CO_2 [5, 47–49], while the USA uses 17% of all its energy to get food through the value chains onto its tables [50]: that is 1000–1500% more energy than what the food itself contains in the form of energy [48, 49].

Food-related health issues, like diabetes in industrialised countries [25], keep growing, as the nutritional value of fruits and vegetables dwindle [51–53] and are less present in conventional farmed produce than in organic produce [54]. Nutritional losses continue to occur during processing and storage, typical for industrialised food systems [55–57].

"The roots of this crisis lie in the preceding decades of excess..." says the global financier George Soros and "...for 25 years the West has been consuming more than we have been producing...living beyond our means" Australian Prime Minister in the Sydney Morning Herald July 25th 2009 [58]. Consumption and growth cannot continue ad infinitum on a finite planet [59]. However, our entire economic model is based on growth funded by debt [60, 61], and as unlikely as economic growth can continue indefinitely, growing debt by civil society, businesses, and government, will also find its limits. A bubble is likely to burst once planetary boundaries and ecosystem services [62, 63], needed to fuel the growth, reach their limits.

2.3 Benefits of agroecology, organic farming and conservation agriculture

CA practiced with LEI, in combination with cover crops, has the ability to harvest atmospheric nutrients, build soil organic matter, increase soil life, loosen soil, increase water holding capacity and aggregate stability, reduce soil compaction, reduce erosion, recharge the water reservoir, improve water quality, reduce nutrient leaching, and increase pest, disease, drought resilience, and CO_2 sequestration [64–71]. A favourable argument in using cover crops as part of CA is the financial viability with increased crop yields and decrease input costs [65, 72, 73].

The yield gap between CvF and organic farming (as another sustainable production system with many parallels to CA), especially with proper diversification practices, is 8–9% smaller than originally estimated [74]. Using 1 ton less synthetic nitrogen fertiliser in organic farming saves the equivalent of 5.2–7.6 tons of carbon emissions [75]. Organic fertiliser in Ethiopia have increased yield by 2–3 times, outperforming inorganic fertilisers [76]. In the US, organic farming achieves comparable yields, but are 28–34% better during droughts [7–9].

Conventional tillage increases soil-based CO_2 respiration and has almost 14 times the carbon emission than no-till [77]; additionally exacerbated by warming global temperatures [78, 79], adding up to 50Pg of soil carbon emissions for the Anthropocene, due to tillage [80]. Organic agriculture uses 2–7 times less energy than conventional agriculture [81, 82] and sequesters 5–15% of greenhouse gas emissions [83, 84]. Tropical agro-forestry systems can sequester between 4 and 6 ton/ha of carbon annually [85–87]. Coghlan [88] even argues that trees planted by local farmers in the Sahel can push back the desert.

A study in Europe has shown that organic farms support more birds, butterflies, beetles, bats and wild flowers than conventional farms, while biodynamic farms have higher levels of soil fertility than organic farms and considerably higher than conventional farms [89, 90].

While we see clear benefits to more sustainable farming practices, such as CA, penetration in South Africa remains low and highly variable between provinces [11], the question remains, why are we not seeing CA products in retail shelves more regularly, or why, when CA poses less risk, do financial institutions not promote CA?

3. Value chain research findings

The findings of this research are concluded from data collected through interviews with some of the largest agricultural VCPs in South Africa, trying to assess if there are blockages inhibiting CA produce from penetrating the market on a broader scale, and if there is potential scalability of CA produce through these value chains. The following narrative details the results of our interviews.

3.1 Banks

The benefits for banks to promote CA to farmers is to end up being less exposed to risk themselves; a capital exposure risk due to drought, potentially exacerbated by climate change. We interviewed four of the largest banks in South Africa, all of whom have been supplying credit to commercial farmers for decades. We asked them whether they had CA tailored products with reduced premiums for farmers because of less perceived risk. Almost all interviewed banks responded in one way or another, saying that they fundamentally did not get involved with production-based decisions around farm practices, such as CA. Three banks argued that these were decisions farmers needed to make for themselves, and as one bank put it, banks would otherwise be in conflict with lender's liability principles.

All of the four banks argued that good production practices for a farmer automatically showed up in production output benefits and a better balance sheet, which in turn would result in a lower risk profile for a farm and in turn, result in a cheaper credit with better premiums. The argument that this might take years for farmers to achieve was generally responded to that that was the nature of farming. One bank confirmed their view that a production method changeover, specifically to CA, would more likely result in an initial increase in cost and reduction in yield, before any yield increases could be observed and benefits would reflect on the balance sheet for farmers to attain better premiums.

Subsequently none of the banks supplied a product that would give farmer credits with reduced premiums should a farmer convert to CA. Only one of the interviewed banks was aware of research that evidenced that CA was a less risky production method, especially in times of drought. Two of the banks stated they would not plan for a specific product for farmers that would entice them to do CA if research were to evidence CA was actually a less risky production method. In contrast the other two banks indicated that they would think about making CA part of the funding application decision or create a product that would have less 'hurdles' during credit approval process, if research showed CA did reduce risk.

All four banks, however, agreed that if CA mitigated risk, it would in any event ultimately reflect on the financial track record and performance over time and subsequently reduce their risk profile, in turn again reducing the premiums these farmers would have to pay. However, a credit offer always remains a decision based on analysing every farm's risk profile, individually.

3.2 Insurers

Insurers are first and foremost exposed to hail and then to drought. Insurers' willingness to take on climate risk on behalf of the farmers makes them also susceptible to the farming practices of the farmer, particularly where new machinery and farming principles such as CA have the ability to reduce drought risk and risk of exposure for insurances. The benefits for insurers to promote CA to farmers is not only about reducing risk of capital exposure to drought, but also other climate change risks and the impact of pests and diseases. With a lower risk premiums insurers charge could be less, which would add economic benefit to the farmers and speed up adoption of CA and a transition to more sustainable farming practices.

We interviewed three of the largest insurers, who together cover around 80–90% of the market in South Africa. None of them had a product tailored to accommodate farmers that farmed with CA practices, or a product that supported the adoption of CA, and none of them indicated that they were thinking of having such a product in future.

When confronted with the questions whether they knew about research that evidenced that CA resulted in more climate resilience and less water stress the insurers argued, similar to banks, that their business model with the way the calculations were done for pricing policy premiums, would automatically benefit those farmers who chose good farming practices that gave consistent yield and had the ability to decrease risk of crop loss at the same time. For example, a farmer that could consistently show stable historic yields, even during draught or 'environmental shocks' would automatically get a cost benefit on the premium of the policy, than a farmer that had bumper yields in good years, but suffered great losses during droughts.

One insurer said, the principle of insurance hinges around good practice, no matter if you use CvF practices or CA practices. Good practice reflects in the historic records, which they would use to price the premium. However, to attain such benefit the track record and historic proof needed to be in place and that would take a few years before reflecting as a better guarantee against crop loss or as a better price on the policy, or both. Farmers who made use of a lot of fertiliser, yet did not look after their soils, this insurer said, might well be likely to show more yield than CA farmers in bumper years, but were also much more likely to suffer greater losses during challenging years. Another insurer said that a well-developed underwriting process would pick up such fluctuation risks and subsequently price more expensively.

When asked whether they as insurer would think about developing products to entice the farmer to take the route of CA and get a better premium without needing to wait many years to benefit, all three interviewed insurers were not thinking of developing such products, nor seemed in favour of it. It would not work with their underwriting principles one insurer argued. One of the few risk mitigating tools they had, this interviewee said, was to work with some form of proof of historical data; you could unfortunately not insure just on a promise that something might happen.

All three of the interviewed insurers knew, or had read something about CA being able to reduce risk by being more climate resilient and building soil structure that would enhance the ability of soil to retain more water. However, to one of the insurers drought was less of a risk than crop loss due to hail and for hail CA had no solution. The interviewee said that they were less exposed to drought, as only after germination would their insurance kick in, and because germination would mitigate a large portion of risk the exposure to drought was less risky than hail. For example, if a farmer has not planted because of low rainfall, or the seed has not emerged because of low rainfall, there will not even be a policy in place to claim against, because the policy is only triggered after the seed has emerged. Because their exposure to drought was so low, getting farmers to farm CA would only have a very small effect on their business in any event, not validating the effort to develop a product targeted at supporting CA.

Another insurer challenged the notion that farmers actually understood the relationship between CA, soil organic matter, the ability to store more soil moisture and what that meant for their crops being more climate resilient. To this insurer, the one farmer he knew only wanted to do CA because of the no-till aspect, which saved fuel and was less capital intensive. This farmer was less thinking about CA benefits on future yields being at less risk due to climate change but was more interested in reducing costs.

3.3 Traders and processors with silos and milling

The interviewed traders and food processors package maize and maize starch products into end consumables that they sell to supermarkets. Consumers' choice in front of retail shelves influences their brand and supporting a procurement of CA produce could attach a sense of sustainability to their brand and grow an awareness amongst their consumers that their brand is ecologically just and fair to the planet's recourse base. The question is if this is a valid and sufficient argument for the traders to get farmers to supply them more CA farmed produce. We interviewed two large traders and food processors in South Africa that were also owned silos and milling operations of maize. Due to their large product profile, their answers considered a wide range of products.

The term CA generally confused both traders as they were not sure how it related to organic and GMO-free farming practices. In contrast to banks that have dedicated agricultural business units with knowledgeable staff, the traders usually purchased produce from other traders and cooperatives, without needing to understand production methods. After explaining CA in a bit more in detail, the interviewee's answers were more cantered around general sustainability including responses around organic and GMO-free produce. The general topic though was still in line with CA principles.

Because the two interviewed traders were buying from other traders, silo owners, and co-operatives, they had no control over what was in the silos from which they attained their maize, or what portion of the maize in a silo was from CA practices. One trader said they would not keep GMO maize separate, even if they had access to it, as they used about 300,000 tons of maize a month and keeping anything separate in their storage, in such type of bulk environments, would not make sense, specifically for South Africa where 70% of maize is farmed as GMO maize. This trader mostly bought from silo owners, other traders and co-ops, and other than during harvest, in order to fill their own silos, they would not buy directly from farmers.

The same trader said that if he had a farmer that farmed using CA, or was GMO-free, it would be just too small for him to go and collect a 100 tons, which is three truck loads, while they are looking for at least, between 3000 and 5000 tons to fill a silo. This trader claimed that such small quantities would not be viable within their system, where they would have to thoroughly clean an entire mill or alternately install a whole new mill for R120m. This miller did not foresee any change happening in the near future for them, and the second trader said he could not honestly comment on whether there was a trend amongst farmers to go GMO-free or CA, as they were too detached from farmers to comment.

Both traders also perceived that there was no demand for CA, GMO-free or organic, and subsequently there was no strategy within their companies to attain certain products or create product ranges that were either GMO-free, organic or farmed with CA practices. One trader confirmed that there was merely demand for GMO-free maize from an insignificant part of the population, a mostly health conscious upper-class society, who at the same time, he criticised, did not understand GMO. There was also no pressure on them from the market side supplied more GMO-free or CA produce. One trader said that they were processing huge volumes and that the odd packet of organic maize, organic flour, non-gluten flour, or GMO-free sold in Woolworths were of such small quantities that they were sourced from completely different channels and producers, rigged to supply such a niche market, which was not theirs to serve.

Both traders agreed that for them as big millers, it was not feasible to separately mill and brand for a potential small volume of GMO-free, CA, or organic demand.

While one trader believed that this situation would not change for them in the near future, the second trader said that they would switch over to GMO-free, once the majority of farmers did so as well and reliable volumes and batches could be processed in that way. However, in contrast to GMO-free, CA was not at all on the radar for them, as the market did not understand what it meant, and the demand was not there.

Around the question of how government could get involved with creating a supportive framework for CA, both responded very similarly in that open market principles of supply and demand should prevail and government should not interfere with legislation or policy. One of the traders said that they did not want a duty on maize, as it existed with wheat, especially where the duty funds disappear and are not invested back into agriculture. This interview participant said that if ever government were to think about a maize duty on such a large staple, it should be used to flow back to agriculture to change the farmers' minds to do CA. One trader responded that government should be careful not to 'play' around with the basic food needs of a nation, and should let free market forces of supply and demand regulate the food supply.

3.4 Supermarkets

Similar to traders, were supermarkets to support CA through preferential procurement could likely attach a sense of sustainability to their brand and grow also an awareness amongst their consumers for being ecologically just and fair to the planet's recourse base. The question arises whether this is a valid and sufficient argument for retailers, who compared to traders have a direct engagement with end consumers, to encourage farmers to use CA.

We interviewed three major retailers in South Africa, all of whom had some form of sustainably farmed produce on their shelves already. The interview was conducted with senior employees of these organisations who were responsible for, or involved with, the purchasing of farm produce. During the interview, the respondents tended to focus their answers more around sustainably farmed products than specifically CA produce. The retailers, as became clear, have not been confronted with CA produce specifically, but usually with a host of differently and sustainably farmed produce, ranging from organic to low carbon etc.

The retailers did not have dedicated shelves that sold CA produce, as most of them had for organic products; they would either be on the same shelf and branded differently or altogether placed somewhere else. None of the retailers had CA farmed maize in their portfolio, and when asked whether they would buy CA produce if it were readily available, two of the three retailers would probably purchase CA products if there were a demand for CA produce. One retailer said it would be pointless to buy it, if the consumer did not understand what it was. To that retailer, the average consumers was more likely to understand, or have heard about GMO-free or organic produce, but not CA produce.

All three retailers agreed that there was very little understanding from the consumer side about CA, and that it was unlikely to change in future. The retailers said that a lot more education would need to take place for the average consumers to understand CA, or even organic farming, and until then, the demand is low and is likely to stay low.

However, one retailer said that it saw CA practices amongst farmers increasing, independent from market demand. They could see it, for example, through produce like sweet corn, with very successful farmers doing no-till sweet corn. To them, the increase in CA seemed likely, and it would be driven from the farmer side, as the benefits of CA were for the farmer and less for the consumer, at least at this stage.

If asked whether they would focus specific product ranges on organic, or sustainably farmed produce, the answers varied between the retailers. Although not specifically focussing on product ranges, one retailer had bad experiences of organic produce and subsequently had more grocery line products like olive oil and biscuits that were organic than fresh produce, which they had tested unsuccessfully a few years earlier. This was related to an inconsistency in supply and price premiums of 25–30% for organic produce which consumers were not prepared to pay. The second retailer said that they also did not have any specific focus on organic or sustainably farmed product ranges, but that they had a wide range of produce and groceries, with a slightly stronger hold on organic fruit and vegetables.

The third retailer had a very high turnover with one specific fruit and because it was as a high-volume product it was fairly easy to maintain the flow of this organic certified product. Through their programme, they said they would try to get as much sustainable produce as they could, and although it was not easy, the whole idea of the programme was to start making farmers think more about how they were farming.

For most retailers, the consumers understood organic farming, but not CA and responded with an unwillingness to start branding another sustainable production method and to educate the consumers. One retailer suggested a softer approach to building a stronger base for sustainability was a better way, than to go out and brand a host of sustainable production methods, it would confuse the consumers.

This retailer also argued that the consumer is often very indifferent to whether produce is farmed in a sustainable manner or not. This retailer argued that their own internal research showed that they could, for example, have tomatoes come from a producer that farmed according to their sustainability programme and another that did not and selling the tomatoes at the same price did not make the consumer chose the sustainably-farmed produce more. The retailer reiterated that it is mostly a benefit to the farmer to farm more sustainably, as many of their suppliers farmed their produce less expensively than those who did not farm sustainably.

4. Discussion

4.1 Summary of data

From both the grounded theory used to analyse the qualitative interview data and the qualitative data collected we have generated **Table 1**. There are four major themes that we could identify using coding principles of grounded theory and **Table 1** shows for each of those themes how each of the VCPs is positioned against a theme.

While some retailers supported organic products, none of the VCP had any form of products which supported CA and with exception to one bank, next to none of the VCP were thinking of or prepared to develop products related to CA. This notion also closely relates to the last column on the right, where all retailers and traders indicated that if they were to choose to support either organic or CA, they would support organic because it is an established brand. In other words, no one was prepared to engage in establishing another brand around CA.

Both retailers and traders also indicated that it is very unlikely that any of their clients would demand CA farmed produce as they were also very unlikely to understand what CA was about in first place. Across the board of all interviewed VCPs there was very little support for CA or to drive a change to more CA farming practices.

	Do you have a CA product?		Do you plan for a future CA product?		Do consumers demand CA?		What are you inclined to support?	
	Yes	No	Yes	No	Yes	No	CA	Organic
Banks (4)	0.0%	100.0%	25.0%	75.0%	n.a.	n.a.	n.a.	n.a.
Insurers (3)	0.0%	100.0%	0.0%	100.0%	n.a.	n.a.	n.a.	n.a.
Traders (2)	0.0%	100.0%	0.0%	100.0%	0.0%	100.0%	0.0%	100.0%
Retailers (3)	0.0%	100.0%	0.0%	100.0%	0.0%	100.0%	0.0%	100.0%
Total	0.0%	100.0%	6.3%	93.8%	0.0%	100.0%	0.0%	100.0%

Table 1.
Summary of responses around CA (sample sizes in brackets).

4.2 Policy discussion

Generally, there was a broad confusion amongst all respondents around what CA meant, and how that related to organics and GMO. Even a retailer that was well versed in the procurement of foods from more sustainable farming practices was not sure what was meant with CA. Many interviewees believed a lot of consumer education efforts would need to happen before they could drive CA from a marketing perspective.

While the retailers did have sustainable products like organic, branding a second sustainability label did not make sense and therefore they also did not put pressure on traders and millers to supply them with CA products. The opinion amongst the retailers and traders was that maize was purchased from other traders, cooperatives and silo owners, where there was no control over the separation of CA maize and non-CA maize. The traders agreed that it would be cumbersome and expensive to try to keep CA maize separate, and it would only ever work if most farmers were to switch over to CA farming.

The feeling from one trader was that generally, government should not get involved in regulating the market, as it did with the duty on wheat. If, however, any duty was to be imposed, the funds from such a duty could be used to fund agriculture and be used to possibly support CA through extension for example. On the discussion of GMO-free products, the traders would not change their operations or invest heavily for a small consumer group perceived to have 'upper-class angst'. The retailers also showed an unwillingness to start branding another sustainable method next to organic produce, which by now only a few consumers understood.

The feeling amongst the banks was that they do not get involved with what is fundamentally a farm production level decision which was to be made by the farmer himself. In any event, lender's liability principles would not allow them to dictate any form of farming operations when giving a credit. Banks argued that if CA mitigated risk, it would automatically show on a farmer's balance sheet, and subsequently affect the risk profile and credibility of the farmer to his benefit; although most respondents agreed that it would take years for a farmer to see such benefits reflecting on his credit profile. While one bank was sure to create a product that in future would assist farmers converting to CA, two more hinted that there was the possibility of a future product that assisted farmers going CA and supported them through a potential initial cash-dip, if CA research proved to be less risky and more productive for farming in the South African context. Policy could therefore support more research into CA benefits for individual regions of South Africa.

All the interviewed insurers seemed to know about research that evidenced that CA resulted in more climate resilience and less water stress. Similar to banks the insurers argued that by the nature of their business model and the way in which the underwriting process works, calculating policy premiums, farmers would automatically benefit if they chose better farming practices that resulted in more consistent yields with decreased risk of crop loss. None of the insurers were also supporting the notion of developing a future product that supported CA. Because mitigating risk is the insurer's business, they would not insure a new system based on a promise that it might mitigate drought in years to come. For one insurer drought risk was a small exposure and therefore drought risk mitigation was for them less high on the agenda.

Because, as argued above, there is an attractive financial argument in favour of CA, farmers are likely inclined to take up CA as a farming practice for their own future financial benefits. With less 'draught risky' farming practices this in turn would position these more favourably in front of financial institutions such as banks and insurers. Yet such a transition would be driven by the farmer and not VCP or policy, in contrast, because almost all VCP showed little interest in developing CA targeting products, a policy driving CA support through the VCPs would likely yield little impact.

Based on our findings a key implication is that government policy concerning CA should endeavour to provide an enabling environment for the future uptake of CA. We suggest that a slow process of change is the route that policy should take, with key aspects focussing on policy enabling training and capacity development of farmers, through field extension and agricultural schools, to adopt CA. Because CA is also a cheaper production system and needs less external inputs, focus should be on smallholder farmers who mostly struggle with access to external inputs.

Over time, a policy that favourably supports CA would grow the farming user base; and in doing so end-consumers would automatically get access on a broader base to more sustainably farmed products without any system change in the value chain, in which the value chain participants have clearly indicated not to drive CA as a system.

5. Conclusion

We have argued that CvF practices have high external input costs and a substantial impact on natural ecosystems, ecosystem services, soil erosion, and CO_2 emissions. These are results of a conventional industrial agricultural complex that also dominates the modern South African agricultural food value chain. We have argued that more sustainable farming practices such as CA are more climate resilient and supply more nutritional value, both of which favourably impact long term food security.

We have raised the question whether South Africa's current food system has the ability to sustain long-term food security and if changes in the existing food value chain complex would be able to drive a transition into a more sustainable and food secure alternative such as CA.

From the research data we can conclude that none of the respondents had a product that supported CA and the general inclination of most interviewed was not particularly in favour to support CA through new product development in their respective institutions. Mostly it was argued that it would either interfere with their specific business-client integrity, or it would not fit into their specific business model or alternatively be too difficult to sell to the end consumer, who understood organic but not CA. The traders argued that with the large volumes and silos they worked in, keeping CA produce separate would be very costly.

For the respondents from the financial institutions CA had the potential to mitigate risk, however in the eyes of most of the respondents there was yet not enough evidence to prove solid risk mitigation. Therefore, they argued, it would be better for them to rely on the existing business model and underwriting process, which would feedback a preferential pricing to a farmer automatically, were he able to use a production system that reduced risk. Subsequently the development of new products to drive the support of CA would not be required. Of the 12 respondents only one indicated that it would develop a product in future that would specifically support the adoption of CA on farmer side. Most retailers and traders indicated that if they were to drive anything sustainable they would support organic which was already an established 'sustainability' brand and needed far less effort to communicate to end consumers.

We can conclude that South Africa's VCPs are neither a support network today, nor will they be one tomorrow; therefore they are not a potential channel to drive a transition. However, while the VCPs were generally supportive of sustainable production methods (such as CA), from an operational perspective and from within their existing business models, VCPs are unlikely become initiators of a sustainable transition driven by CA supportive products. For policy purposes we may deduct that efforts for a transition and required training of farmers would need to focus on education rather than enforcing policy on value chains and their existing business models.

Author details

Wolfgang Johann von Loeper[1*], Scott Drimie[1] and James Blignaut[1,2]

1 School of Public Leadership, Stellenbosch University, Stellenbosch, South Africa

2 South African Environmental Observation Network, Pretoria, South Africa

*Address all correspondence to: wolfgang@mysmart.farm

References

[1] Hui C. Carrying Capacity of the Environment. International Encyclopedia of the Social & Behavioral Sciences. 2015;10(2):155-160

[2] Rockström J, Steffen W, Noone K, Persson Å, Chapin FS, Lambin EF, et al. Planetrary boundaries: Exploring the safe operating space for humanity. Ecology and Society [Internet]. 2009;1-36. Available from: http://medcontent.metapress.com/index/A65RM03P4874243N.pdf [cited 2014 Feb 7]

[3] Steffen W, Richardson K, Rockström J, Cornell SE, Fetzer I, Bennett EM, et al. Planetary boundaries: Guiding changing planet. Science (80-.). 2015;**347**. DOI: 10.1126/science.1259855

[4] De Schutter O, Vanloqueren G. The New Green Revolution: How twenty-first-century science can feed the world. Solution Journal. 2011;**2**:33-44

[5] Smith P, Bustamante M, Ahammad H, Clark H, Dong H, Haberl H. Chapter 11: Agriculture, forestry and other land use (AFOLU). In: Intergovernmental Panel on Climate Change. Berlin: Final Draft; 2014

[6] Nelson GC, Rosegrant MW, Koo J, Robertson R, Sulser T, Zhu T, et al. Climate change: Impact on agriculture and costs of adaptation. Vol. 21. Washington, DC: International Food Policy Research Institute; 2009

[7] Delate K, Cambardella C. Organic production: Agroecosystem performance during transition to certified organic grain production. Agronomy Journal. 2004;**96**:1288-1298

[8] Lim L. Organic agriculture fights back. Science Society. 2002;**16**:30-32

[9] Rodale Institute. Regenerative Organic Agriculture and Climate Change: A Down-to-Earth Solution to Global Warming. Vol. 1400. Kutztown: Rodale Institute; 2014

[10] Blignaut JN, De Wit MP, Midgley S, Crookes DC, Knot J, Drimie S, et al. Sustainable agriculture: A viable option for enhanced food and nutritional security and a sustainable productive resource base in South Africa: An investigation. Pretoria: Green Fund; 2014

[11] Blignaut J, Knot J, Smith H, Nkambule N, Crookes D, Saki A, et al. Promoting and Advancing the Uptake of Sustainable, Regenerative, Conservation Agricultural Practices in South Africa with a Specific Focus on Dryland Maize and Extensive Beef Production Key Findings and Recommendations. Pretoria: Green Fund; 2015. p. 1-19

[12] Pittelkow CM, Liang X, Linquist BA, van Groenigen KJ, Lee J, Lundy ME, et al. Productivity limits and potentials of the principles of conservation agriculture. [Internet]. Nature. 22 Oct 2014;000(00):1-6. Available from: http://www.nature.com/doifinder/10.1038/nature13809 [Cited 2014 Oct 22]

[13] Midgley S, Drimie S, von Loeper WJ, Mudavanhu S, Blignaut J, Smith H, et al. Implications for a Future Agrarian Structure in South Africa based on Conservation and Sustainable Agriculture: Alignment of a Multi-institutional and Multi-policy Landscape. Pretoria: Green Fund; 2015. p. 1-21

[14] SA Department of Agriculture Forestry and Fisheries. South African Integrated Growth and Development Plan 2012. Pretoria: Department of Agriculture, Forestry and Fisheries; 2012

[15] SA Department of Agriculture Forestry and Fisheries. South Africa

Agricultural Policy Action Plan. Pretoria: SA Department of Agriculture Forestry and Fisheries; 2014

[16] FAO. Food Security [Internet]. FAO Policy Briefs. New York City; 2006. Available from: http://www.fao.org/forestry/13128-0e6f36f27e0091055bec28ebe830f46b3.pdf

[17] Rosen C. Chapter 3: Living in Ecosystems. Fraying the web of life. In: World Resources 2000-2001, Elsevier; 2000. p. 147-224

[18] Murray W. Save and Grow in Practice: Maize Rice Wheat, A Guide to Sustainable Cereal Production. Rome: Food and Agriculture Orgvanization of the United Nations (FAO); 2016. DOI: 10.1017/CBO9781107415324.004

[19] DAFF. Crops and markets. South African Dep Agric For Fish. 2008;89(937):1-34

[20] Pettigrew SF. Ethnography and Grounded Theory: A Happy Marriage?. http://www.acrwebsite.org [Internet]. 2000. Available from: http://www.acrwebsite.org/search/view-conference-proceedings.aspx?Id=8400 [cited 2016 Aug 21]

[21] Savin-Baden M, Major CH. Qualitative Research: The Essential Guide to Theory and Practice. London and New York: Routledge; 2012

[22] Gibbs. Core Emlements Part 2. Grounded Theory [Internet]. Available from: https://www.youtube.com/watch?v=dbntk_xeLHA&feature=related [Cited 2016 Aug 21].

[23] Glaser BG. The Grounded Theory Methodology of Barney G. Glaser, Ph.D—What is GT? Grounded Theory Institute; 2014. Available from: http://www.groundedtheory.com/what-is-gt.aspx [Accessed: August 21, 2014]

[24] Pengue W, Brien MO, Garcia F, Sims R, Robert W, Kauppi L, et al. Assessign Global Land Use: Balancing Consumption with Sustainble Supply. Paris: UNEP—International Resource Panel; 2013

[25] Pretty J. Agricultural sustainability: Concepts, principles and evidence. Philosophical Transactions. Royal Society of London. 2008;**363**:447-465. DOI: 10.1098/rstb.2007.2163

[26] Robertson GP, Vitousek PM. Nitrogen in agriculture: Balancing the cost of an essential resource. Annual Review of Environment and Resources. 2009;**34**:97-125. DOI: 10.1146/annurev.environ.032108.105046

[27] Ewers RM, Didham RK. Pervasive impact of large-scale edge effects on a beetle community. Proceedings of the National Academy of Sciences. 2008;**105**:4-7

[28] Pingali PL, Pardey PG. Reassessing International Agriculture Research for Food and Agriculture. Montpellier: Global Conference on Agricultural Research for Development; 2010

[29] Letourneau DK, Armbrecht I, Rivera BS, Lerma JM, Carmona EJ, Daza MC, et al. Does plant diversity benefit agroecosystems? A synthetic review. Ecological Applications. 2011;**21**:9-21

[30] Bernstein H. Food Sovereignty: A Sceptical View Food 2013. Available from: http://www.iss.nl/fileadmin/ASSETS/iss/Research_and_projects/Research_networks/ICAS/ICAS_Review_Paper_4_Bernstein.pdf [Accessed: April 4, 2014]

[31] Lithourgidis AS, Dordas CA, Damalas CA, Vlachostergios DN. Review article annual intercrops: An alternative pathway for sustainable agriculture. Australian Journal of Crop Science. 2011;**5**:396-410

[32] Diaz RJ, Rosenberg R. Spreading dead zones and consequences for marine ecosystems. Science. 2008;**321**:926-929. DOI: 10.1126/science.1156401

[33] Rohr JR, Schotthoefer AM, Raffel TR, Carrick HJ, Halstead N, Hoverman JT, et al. Agrochemicals increase trematode infections in a declining amphibian species. Nature. 2008;**455**:1235-1239. DOI: 10.1038/nature07281

[34] Daly GL, Lei YD, Castillo LE, Muir DC, Castillio LE, Wania F. Accumulation of current-use pesticides in neotropical montane forests. Environmental Science & Technology. 2007;**41**:1118-1123

[35] Hemingway J, Field L, Vontas J. An overview of insecticide resistance. Science Magazine. 2002;**298**:96-97

[36] Brent KJ, Hollomon DW. Fungicide Resistance: The Assessment of Risk. Brussels: Fungicide Resistance Action Committee; 2007

[37] Heinemann JA, Massaro M, Coray DS, Agapito-Tenfen SZ, Wen JD. Sustainability and innovation in staple crop production in the US Midwest. International Journal of Agricultural Sustainability. 2014;**12**: 71-88. DOI: 10.1080/14735903.2013.806408

[38] Pimentel D, Harvey C, Resosudarmo P, Sinclair K, Kurz D, McNair M, et al. Environmental and economic costs of soil erosion and conservation benefits. Science (80-). 1995;**267**:1117-1123

[39] Reusser L, Bierman P, Rood D. Quantifying human impacts on rates of erosion and sediment transport at a landscape scale. Geology. 2015:1-4. DOI: 10.1130/G36272.1

[40] Paoletti M, Pimentel D. Environmental risks of pesticides versus genetic engineering for agricultural pest control. Journal of Agricultural and Environmental Ethics. 2000;**12**:279-303. DOI: 10.1023/A:1009571131089

[41] UNEP. Towards a Green Economy: Pathways to Sustainable Development and Poverty Eradication. St-Martin-Bellevue, France: United Nations Environment Programme; 2011

[42] Matson PA, Vitousek PM. Agricultural intensification: Will land spared from farming be land spared for nature? Conservation Biology. 2006;**20**:709-710. DOI: 10.1111/j.1523-1739.2006.00442.x

[43] Fischer J, Brosi B, Daily GC, Ehrlich PR, Goldman R, Goldstein J, et al. Should agricultural policies encourage land sparing or wildlife-friendly farming? Frontiers in Ecology and the Environment. 2008;**6**:380-385. DOI: 10.1890/070019

[44] Perfecto I, Vandermeer J. Biodiversity conservation in tropical agroecosystems: A new conservation paradigm. Annals of the New York Academy of Sciences. 2008;**1134**:173-200. DOI: 10.1196/annals.1439.011

[45] FAOSTAT. Global Distribution of Agricultural Land. UN: FAO; 2014. pp. 1-10. Available from: http://faostat3.fao.org/faostat-gateway/go/to/home/E [Accessed: April 20, 2005]

[46] Baleta H, Pegram G. Water as an input in the food value chain. In: Understanding the Food Energy Water Nexus. South Africa: Wolrd Wildlife Fund—South Africa; 2014

[47] Adhya T, Sharma P, Gogoi KA. Mitigating greenhouse gas emission from agriculture. In: Climate Change and Crops. Berlin, Heidelberg: Springer-Verlag; 2009. pp. 329-344. DOI: 10.1007/978-3-540-88246-6

[48] Cleveland C, Kaufmann R. Energy and Resource Quality. New York: Wiley Interscience; 1986

[49] June S, Günther F. Vulnerability in agriculture: Energy use, structure and energy futures. In: Pap. Present. INES Conf. KTH Stock. 2000

[50] Pimental D, Armstrong L, Flass C, Hopf F, Landy R, Pimentel M. Interdependence of food and natural resources. In: Food and Natural Resources. San Diego: Academic Press; 1989

[51] Rodale Institute. Nutritional value of fruits, veggies is dwindling—Health-Diet and nutrition—msnbc.com. Rodale Inst.; 2012. Available from: http://www.nbcnews.com/id/37396355/ns/health-diet_and_nutrition/t/nutritional-value-fruits-veggies-dwindling/#.U4SdkXKSySp [Accessed: May 27, 2014]

[52] Mitchell AE, Hong Y-J, Koh E, Barrett DM, Bryant DE, Denison RF, et al. Ten year comparison of the influence of organic and conventional crop on flavonoids. Journal of Agricultural and Food Chemistry. 2007;**55**:6154-6159. DOI: 10.1021/jf070344+

[53] Huber M, Rembiałkowska E, Średnicka D, Bügel S, van de Vijver LPL. Organic food and impact on human health: Assessing the status quo and prospects of research. NJAS—Wageningen Journal of Life Sciences. 2011;**58**:103-109. DOI: 10.1016/j.njas.2011.01.004

[54] Volakakis N, Seal C, Sanderson R, Dominika S, Stewart GB, Benbrook C, et al. Higher antioxidant and lower cadmium concentrations and lower incidence of pesticide residues in organically grown crops: A systematic literature review and meta-analyses. The British Journal of Nutrition. 2014:1-18. DOI: 10.1017/S0007114514001366

[55] Labuza TP, Tannenbaum SR. Nutrient losses during drying and storage of dehydrated foods. Critical Reviews in Food Science and Nutrition. 1972;**3**:217-240

[56] Reddy MB, Love M. The impact of food processing on the nutritional quality of vitamins and minerals. In: Impact Process. Food Saf. US: Springer; 1999. pp. 99-106

[57] Giannakourou MC, Taoukis PS. Kinetic modelling of vitamin C loss in frozen green vegetables under variable storage conditions. Food Chemistry. 2003;**83**:33-41. DOI: 10.1016/S0308-8146(03)00033-5

[58] Jehu. More Notes on the Deleveraging Economy. Pogo Chron; 2009. Available from: https://pogoprinciple.wordpress.com/2009/08/18/ [Accessed August 21, 2016]

[59] Heinberg R. Temporary Recession or the End of Growth? 2009. Available from: www.richardheinberg.com [Accessed: June 1, 2015]

[60] Rubin J. The end of growth: But is that all bad. Toronto: The Globe and Mail; 2012

[61] Korowicz D. Tipping Point: Near-Term systemic implications of a Peak in Global Oil production. Dublin: Feasta Foundation for the Economics of Sustainability; 2010

[62] Blewitt J. Understanding Sustainable Development. London: Earthscan; 2008

[63] Pretty J. The consumption of a finite planet: Well-being, convergence, divergence and the nascent green economy. Environmental and Resource Economics. 2013;**55**:1-25. DOI: 10.1007/s10640-013-9680-9

[64] Bunch R. Restoring the Soil: A Guide for Using Green Manure/Cover Crops to Improve the Food Security of Smallholder Farmers. Winnipeg: Canadian Foodgrains Bank; 2012

[65] Bryant L, Stockwell R, White T. Counting Cover Crops. Washington DC: National Wildlife Federation; 2013

[66] Feeser J, Zinati G, Moyer J. Beyond Black Plastic: Cover Crops and Organic No-till for Vegetable Production. Kutztown PA: Rodale Institute; 2014

[67] Bowie MH, Wratten SD, White AJ. Agronomy and phenology of companion plants of potential for enhancement of insect biological control. New Zealand Journal of Crop and Horticultural Science. 1995;**23**:423-427. DOI: 10.1080/01140671.1995.9513919

[68] Kumar S, Bahl JR, Bansal RP, Gupta AK, Singh V, Sharma S. High economic returns from companion and relay cropping of bread wheat and menthol mint. Industrial Crops and Products. 2002;**15**:103-114

[69] Findlay N. Cover crops: What are they and why are they used. Durban: KZN Department of Agriculture; 2011. p. 1-3

[70] Bezuidenhout SR. The Influence of Two Cover Crop Species on the Growth of *Zea mays* and *Cyperus*. Pretoria: University of Pretoria; 2012. p. 1-8

[71] Vandermeer JH. The Ecology of Intercropping. Cambridge: Cambridge University Press; 1989

[72] Strauss J. Wheat yield and crop rotation. Stellenbosch: Western Cape Department of Agriculture; p. 1-3

[73] Hardy M, Strauss J, Laubscher S. Gross Margin Analysis of Crop and Crop/Pasture Rotation Systems that are being Evaluated in the Long-term Crop Rotation Trial at the Langgewens Agricultural Research Farm—2002 to 2010. Vol. 1. Stellenbosch: Department of Agriculture, Western Cape Institute for Plant Production; 2011

[74] Ponisio LC, M'Gonigle LK, Mace KC, Palomino J, de Valpine P, Kremen C. Diversification practices reduce organic to conventional yield gap. Proceedings of the Royal Society B: Biological Sciences. 2014;**282**:20141396-20141396. DOI: 10.1098/rspb.2014.1396

[75] Wood S, Cowie A. A review of greenhouse gas emission factors for fertiliser production. 2004

[76] Edwards S. Greening Ethiopia series. Science Society. 2004;**23**:4-8

[77] Reicosky DC. Tillage-induced CO_2 emission from soil. Nutrient Cycling in Agroecosystems. 1997;**49**:273-285

[78] Raich J, Schelsinger W. The global carbon dioxide flux in soil respiration and its relationship to vegetation and climate. Tellus. 1992;**44B**:81-99

[79] Schlesinger W, Andrews J. Soil respiration and the global carbon cycle. Biogeochemistry. 2000;**48**:7-20. DOI: 10.1023/A:1006247623877

[80] Topp CFE, Wang W, Cloy JM, Rees RM, Hughes G. Emissions from agricultural soils. Biochemistry. 2013;**48**:972-987. DOI: 10.3390/e15030972

[81] Medard G. Empty breadbasket? The coming challenge to America's food supply and what we can do about it. 1981

[82] Cormack B, Metcalfe P. Energy Use in Organic Farming Systems. London: Ministry of Agriculture, Fisheries and Food; 2000

[83] Lal R. Soil carbon sequestration impacts on global climate change and food security. Science (80-). 2004;**304**:1623-1627

[84] Lal R, Negassa W, Lorenz K. Carbon sequestration in soil. Current Opinion in Environment Sustainability. 2015;**15**:79-86. DOI: 10.1016/j.cosust.2015.09.002

[85] Lasco R, Sales R, Extrella R, Saplaco S, Castillo A, Cruz R, et al. Carbon

stock assessment of two agroforestry systems in a tropical forest reserve in the Philippines. Philippine Agriculturist. 2001;**84**:401-407

[86] Schroeder P. Agroforestry systems: Integrated land use to store and conserve carbon. Climate Research. 1991;**3**:53-60

[87] Hepperly P. Organic farming sequesters atmospheric carbon and nutrients in soils. 2003. Available from: http://www.newfarm.org/depts/NFfield_trials/1003/carbonwhitepaper.shtml [Accessed: November 1, 2014]

[88] Coghlan A. More crops for Africa as trees reclaim the desert. New Scientist (1971). 2006;**192**:11

[89] Zaller J, Koepke U. Effects of traditional and biodynamic farmyard manure amendment on yields, soil chemical, biochemical and biological properties in a long-term field experiment. Biology and Fertility of Soils. 2004;**40**:222-229. DOI: 10.1007/s00374-004-0772-0

[90] Kennedy AC, Reganold JP, YA L, Carpenter-Boggs L. Organic and biodynamic management: Effects on soil biology. Soil Science. 2000:1651-1659

Chapter 6

Coastal Community Adaptation to Climate Change-Induced Salinity Intrusion in Bangladesh

Golam Rabbani, Sirazoom Munira and Samia Saif

Abstract

Bangladesh, a country with a size of 147,570 km^2, has the largest delta in the world with one-third of the country residing a costal shoreline. The livelihood of the inhabitants is exposed to a series of tribulations causing radical setbacks, of which natural disasters like tropical cyclones bring large-scale salinity intrusion. Occurring due to three main causal factors including climate change, sedimentation, and low water flow, salinity poses challenges in agriculture and overall food security, hence impeding the health and livelihood of marginalized women, children, and elderly and explicably the overall vulnerable population at large. Against that backdrop, this chapter will delineate three broad spectra of Bangladesh's approach to "living with salt": (1) the assessment of salt intrusion in water and soil in Bangladesh's coastal zones due to climate change, (2) understanding the vulnerabilities to salinity within marginalized population, and (3) coping or adapting strategies to combat and live with salt. The chapter also includes the findings of a recent case study conducted on Bangladesh coastal zone to demonstrate the current livelihood conditions under salinity, conditions of the actions taken by the government and nongovernment organizations, gaps, and recommendations for a more resilient coastal community.

Keywords: coastal salinity, climate change, adaptation, Bangladesh, resilience

1. Introduction

Bangladesh is one of the most vulnerable countries due to climate change. According to the Global Climate Risk Index 2017, Bangladesh ranked sixth as the most vulnerable country and as the worst affected country by extreme weather. Climate change has increased the frequency and intensity of natural disasters such as cyclone, storm surge, flood, etc. The long-term impacts of climate change are temperature rise due to global warming, sea-level rise, salinity intrusion, drought, heat waves, cold waves, etc. The coastal belt of Bangladesh is severely impacted by salinity intrusion. In Bangladesh, salinity affected 83.3 million hectares of land in 1973 which increased to 102 million hectares in 2000. In 2009, the amount has increased to 105.6 million hectares. Over the last 35 years, salinity has increased to 26% within the country [1].

Bangladesh has a geographically and hydrologically disadvantaged setting and is highly vulnerable to sea-level rise and other related natural hazards, with major impacts on women, children, and food and energy securities [6]. All these multiple vulnerabilities and climate change impacts are enhancing poverty, migration, and inequality in the country [7]. The government of Bangladesh has been able to identify different eco-zones which are vulnerable to climate change in its own distinct ways. These zones include southwest and central coast of Bangladesh including Bagerhat, Satkhira, Barguna, Patuakhali, and Bhola, which are the major affected coastal districts in the country.

It may as well be noted that IPCC Fourth and Fifth Assessment Report identified a number of climate-sensitive diseases which are waterborne (e.g., diarrhea and cholera). The Government of Bangladesh has been trying through its initiatives and own national funds to ensure that the vulnerable, socially excluded, disadvantaged households are supported with proper water, sanitation and hygiene (WASH) facilities. A number of disasters including flood, cyclone, river bank erosion, waterlogging, landslide, etc. have already hit the country in the past and posed themselves as critical disasters affecting the country's stability in the face of climate change. These disasters affect the entire environmental settings, infrastructure, and natural resource base which are essential for local livelihoods. Cyclone and storm surge, salinity intrusion, coastal flooding, waterlogging, variation in temperature, and rainfall jeopardize coastal livelihoods [8]. The intrusion of excess salinity has pushed the people living in the coastal zones to several dilemmas [9]. However, coastal communities have emerged with their local-level adaptation techniques, being able to set example across the world to "live with salt."

The following chapter summarizes the scenario of climate change induced salinity intrusion and impacts on livelihood communities. It also presents some insights from a research project conducted by Bangladesh Center for Advanced Studies (BCAS) which conducted a case study to understand coastal salinity in Bangladesh. As such, the chapter will provide a narrative on the current problems, present initiatives by the community to address the salinity challenges and way forward.

2. Understanding the vulnerabilities to salinity within marginalized population

2.1 Salinity intrusion and impacts on livelihoods

The salinity intrusion in the coastal belt of Bangladesh is severely affecting the life and livelihood of the community people. People are suffering for pure drinking water, irrigation water, and loss of agricultural lands. The salinity in the river water also changes the aquatic ecosystem and fishery sector. The average share of agricultural production in the national GDP is 35.53% [3]. As soil salinity increases each year, dry season is only favorable for the cultivation of crops in the coastal region. However, dry-season agriculture is becoming difficult due to salinity intrusion in the coastal belt of Bangladesh. Each year this is negatively impacting the crop production across the coastal belt [10]. Since coastal salinity poses great challenges on agriculture, community people are sometimes forced to convert the agricultural land to shrimp fields and other alternative forms of livelihood earning. The salinity also created shortage of grazing land and fodder for the livestock. This is affecting not only the livelihood of the community people but also the health and growth of the children in the area. Due to lack of pure drinking water, people especially the girls have to travel long distance to collect water. These cause dropout from school and sexual harassment. Moreover, intake of salt water causes high blood pressure among the young adults in the coastal Bangladesh [5].

2.2 Impacts of climate change within coastal communities

In Bangladesh about 80% of the land is floodplains with a low mean elevation above the sea level. The average elevation of the southwest coastal zone ranges from 1 to 2 m and in the southeast coastal zone 4–5 m [2]. The geographical location and flat topography of the country make the country more vulnerable to sea-level rise. Climate change has also increased the frequency and intensity of natural disasters. There is evidence of a 5–10% increase in intensity (wind speed) that would contribute to enhanced cyclone storm. Increased intensity of cyclone damages coastal infrastructure including roads, water supply, sanitation systems, administrative buildings, and cyclone shelters.

A devastating cyclone Sidr lashed the coastal region of Bangladesh on November 15, 2007. According to the government, it caused 3363 human deaths, and damage to property, livestock, and crops was estimated to be USD 1.7 billion. Due to severe salinity caused by cyclone Sidr, around 5 lakh hectares of cultivable land in the coastal region have remained uncultivated for the last 10 years [6]. According to the Soil Science Institute, around 4 lakh hectares, out of a total 7.41 lakh hectares of arable land, are highly contaminated by saline water. The farmers of the region are unable to cultivate their land due to the high level of salinity, as much as 16–18 deci-semen per cubic meter water, whereas the tolerable limit is 4–8 deci-semen per cubic meter to cultivate rice [6]. According to the Department of Agricultural Extension due to salinity caused by cyclone Sidr, rice production has decreased by around 4 lakh tons over the last 5 years. Saline water is continuously entering cultivable land through the faulty embankments damaged by cyclone Sidr.

Another devastating cyclone, cyclone Aila, hits the coast of Bangladesh on May 29, 2009. Cyclone Aila dewatered about 10% of the ponds, and 90% of the tube wells were submerged. Women spent an average of 4–5 hours a day collecting water, often walking 2–3 km to reach the nearest safe water sources or collection point [5]. People drank unsafe water or spent their limited financial resources on the collection of safe water or purchasing drinking water. After the cyclone Aila, the surface water and also groundwater of the coastal area of Bangladesh became saltier making drinking water as one of the key challenges in the area [4].

Bangladesh has a coastline of 710 km, and the coastal zone extends over 47,150 km^2 areas. The coastal zone has a population of 38.52 million (BBS 2011). About 20 million people in the coastal areas of Bangladesh are affected by salinity for pure drinking water.

3. Materials and methods

The Bangladesh Centre for Advanced Studies (BCAS) in its recent study conducted a research to understand the context-specific, ecosystem-friendly, and climate-adaptive agricultural and livelihood practices in three vulnerable zones of Bangladesh, which includes coastal, Char, and hilly regions. The study used a mixed-method approach by using both quantitative and qualitative methodological techniques to meet its objectives. For this purpose, national-level policy documents were consulted to understand the gap between existing knowledge and the scenario in the locality at present.

The BCAS research team conducted this study by incorporating focus group discussions (FGDs), key informants interview (KIIs) and household survey using semi-structured questionnaire from the communities. Data collected included demographical information, community's socio-economic condition, resilience planning and disaster preparedness, access to basic services, alternative livelihoods, people's knowledge and awareness on climate change impacts, options on community-based adaptation options indifferent sectors. The questionnaire also included questions to assess how much capacity building is needed for the communities, local government institutes and other stakeholders including the NGOs, CSOs and CBO communities. The data set was made to be disaggregated (where appropriate) by location, gender, ethnicity, and disability. To collect data and information regarding impacts of climate change, community adaptation options, DRM planning, and other key relevant issues, a checklist was developed to conduct the focus group discussions (FGDs). The FGDs were conducted with mixed group (men and women), and the other was with women participants only. Although the research was conducted using the expertise of a skilled team, it had some inherent limitations. The research was conducted in a number of selected villages of the three eco-vulnerable regions of Bangladesh. Hence, the villages represent a fractional picture of the real scenario and corresponding problems in the area. The respondents of the study were to rely on data which were not only just based on perception and observation but also on the informants' reminiscence of past extreme events and ways of tackling.

Appropriate sampling technique and right size of sample for the best estimation of the population characteristics were two important factors for designing the sampling. Determination of representative sample size is a challenging issue. In most cases, the previous information required to determine the sample size particularly standard deviation of population is not available. The alternative way is to consider $p = q = 0.50$ that allows maximum standard error and also ensure maximum size of sample for specific requirements. The sample size of the baseline study has been estimated in such a way that the obtained result of the study may be within 5% of the true value with 95% confidence interval.

4. Results and discussion

In Bangladesh, agriculture is a central livelihood activity. The people living in the study region, i.e., the coastal zone, blame salinity in the water, too much or too little water, to be harmful for agricultural productivity, but they also have some adaptation strategy and mechanisms to combat these problems. For instance, they have been introduced to stress-tolerant types of agricultural crop varieties to adapt to climatic conditions. The people receive training on various areas including homestead gardening, integrated farm management, integrated pest management, compost preparation, animal manure, and crop rotation management. However, 77% of the respondents from the coastal zone stated that they did not receive any training on agricultural practices, which would be quite useful otherwise. To tackle food security issues, respondents from the coastal zones stated that they depend on less expensive food, they take help from friends and relatives, they intake less than required portions, and they also sometimes go for microcredit when there is food crisis.

The BCAS study also helped to understand the condition of the coastal community at crises and their subsequent shelter stories. The study shows that the community people's access to shelters have a connection with the time of receiving the warning signals. In addition, difficulties in getting early warning message and communication are also reasons for delayed responses to evacuate. The condition of water and sanitation practices of the people in the coastal zone show that the condition of hygiene becomes questionable during extreme events. Although people in the coastal zones have access to a range of sources of drinking water which include deep tube well, shallow tube well, rainwater harvesting (RWH), pond sand filter, ponds/canals and streams, etc., they still have to keep reserves when natural disasters take place. On the other hand, in terms of decision-making at the household level, it can be seen that in the coastal zones, women are more engaged in some of the key decisions. Both government and nongovernment organizations have initiated awareness programs on the importance of integration of women in household- and community-level decisions. As one of the key results of the study, it confirmed the essence of resilience of the region to changing conditions brought about by climate change.

In Bangladesh, the preparedness to disasters is subject to the availability of proper and correct signals. Early warning signal messages in Bangladesh are provided prior to any extreme event, but some people still do not move from home. Early warning system (EWS) needs to be strengthened and is essential to make sure that each and every person follows the early warning messages. This would complement the adaptive capacities of the coastal communities to address climate adversities with a sophisticated lead time. In Bangladesh, way to safe, accessible and affordable water and sanitation is still hindered by a number of factors. Under the current situation of Bangladesh's coastal zones, people highly benefitted from the various training programs that are conducted to address various climate components to the communities. It can be also assessed that a large number of people help from the training programs that capacitate the communities in critical ways that help to overcome disaster aftermath. Hence, people's expectation to receive adequate training on agriculture and disaster management has been increased at present due to its effectiveness. The needs to have agricultural training should be met by the government to understand the best practices in achieving the highest output for crops, both before and after the climatic disasters. In addition to agricultural training, livelihood training and awareness programs should be thoroughly and regularly conducted to increase the resilience of people at large. Responsible authorities should emphasize on fostering people's knowledge on climate change through knowledge platforms, training, and awareness programs.

The study investigated which route was taken by the households while collecting water for household activities and by whom was the water collected. In all three study locations, it was observed that females bear the major responsibility for collecting household water. Many people from the coastal areas mentioned that they use the neighbor's toilet, the use of which becomes increasingly difficult during the onsets of natural disasters. This is particularly true for women, because of the conservative environment in the country.

During monsoon when there is heavy rainfall, the conditions of the latrines in the coastal zone become even worse. This leads to unsafe and unhygienic conditions within the sanitation facilities, with worsened situation during excessive rainfall which causes waterlogging. During most disasters, people in the coastal zones mentioned through this study that their latrines become dysfunctional. They expect help and support to be extended from the government and nongovernmental organizations to follow up on these conditions. The Bangladesh government is aware that they should address this with serious consideration to prevent diseases and facilitate ease in access, especially for disabled and elderly people. In the coastal zones, some people still go to the jungles and bushes for defecation purposes because of a lack of proper facility installed close to the house. Also, the distance of the latrines is another important concern, especially for the elderly people who cannot walk to the toilets. People in all the regions stated that there is a problem of contamination in the drinking water when natural disasters hit. According to the study, cyclones have been attributed to cause problems in their source of drinking water as mentioned by 81% of the coastal community respondents. On the other hand, storm surge was mentioned by 74% of the respondents, excessive rainfall by 36%, and salinity intrusion in water resources by 23%. This shows that people are having a clear perception of the disaster types and how they are responsible for contaminating the water that they use for drinking and cooking. The following information was cross-checked through FGD, and according to the discussion, the major sources of drinking water are mainly affected by cyclone in the coastal zones. People in the coastal zones have indicated that their area is largely hit by cyclone and storm surge. It appears evident through the study's FGDs and KIIs that salinity induced by cyclones and storm surge is one of the biggest problems in the coastal zone. These are positions where the community people expect the use of better technologies and support that can help provide fresh drinking water. Many NGOs are already working in the coastal zone to improve the quality of drinking water. However, climate-resilient water supply and sanitation technologies would help the local communities for long-term sustainability.

4.1 Quantitative analysis from the aforementioned field survey

The field survey that was conducted was made comprehensive through the inclusion of interviews and focus group discussions. Some quantitative information from the survey helped to understand the basic concerns of the coastal communities. The following section shows some of the quantitative data that was collected through the survey analyses that explain the impacts of hazards on the livelihoods of the coastal community people.

A team of the BCAS went to the coastal area of Bangladesh where they interviewed a number of people living and dealing with salinity in the coastal zone. Among them, 38% of the respondents were male, and 68% of the respondents were female (**Figure 1**). The informants, cooperative and interactive, mentioned about a number of their problems and issues that they face living in the coastal zones of Bangladesh, which has implications on their livelihood, lifestyle, agriculture, and health aspect.

4.2 Current livelihood options in the study location

Figure 2 demonstrates the percentage of the respondents and their distribution of the livelihood options in the studied areas within the coastal zone. About 35% of the

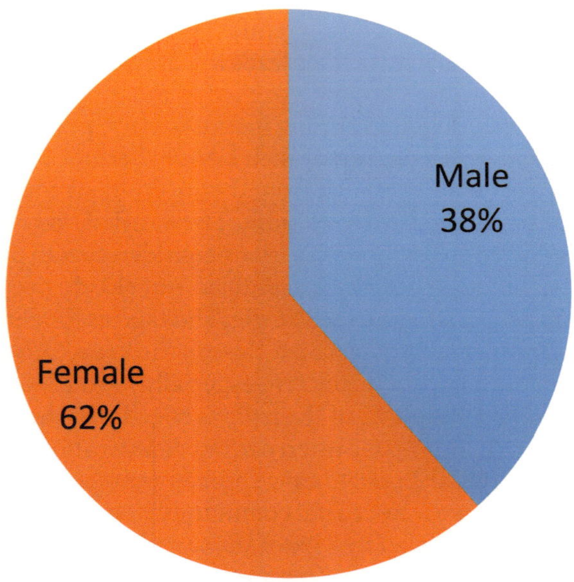

Figure 1.
Percentage distribution of the respondents in the coastal districts during the survey.

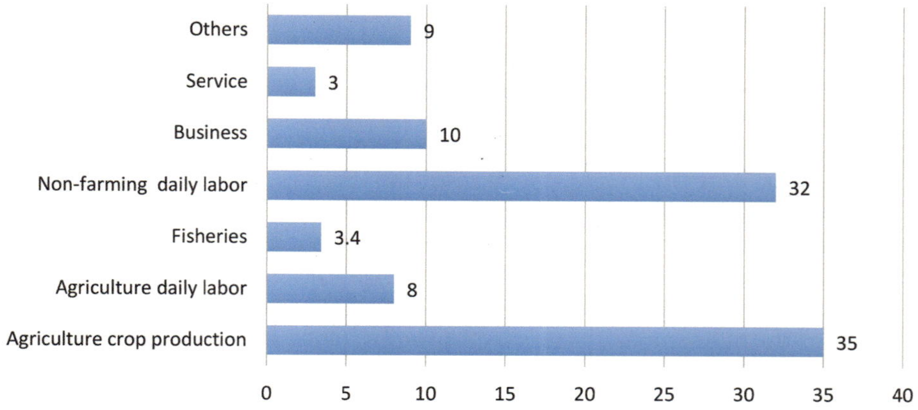

Figure 2.
Percentage of respondents and their distribution of the livelihood options in the study districts of the coastal zone.

people were found to be associated with agricultural crop production. They make their living on producing crops for the benefits of themselves and the community they live in. However, only 3% of the people are service holders. They are the minority in the community of the coastal zones. About 10% of the people make their living running small businesses like grocery shops, running vehicle businesses, etc. Among the second majority, which is 32% of the people, are related to nonfamily daily labor activities. There is an extensive demand on fisheries for supplying fish and food sources; hence, the life of 3.4% of the people depends upon fisheries.

4.3 Sources of drinking water

Figure 3 demonstrates the sources of drinking water in the study areas. From the study, it was found that majority of the people rely on deep tube well, which amounts to 62%, since the groundwater is believed to be pure and risk-free. Nearly 21% of people rely on direct rainwater which they store by means of tank pond and other resources and use them whenever necessary. A minority of the people, which is 5%, depends on river canals. However, due to hygiene concerns, the ones who are in need and do not have sufficient resources for drinking water only rely on canal water. Nearly 23% of the people depend on ponds and canals which they use as household water. About 10% of the people depend on pond sand filters as a source of drinking water. The least number of people can afford rainwater harvesting system (RWHS) which is 2% and 9% of the people who depend on shallow tube well.

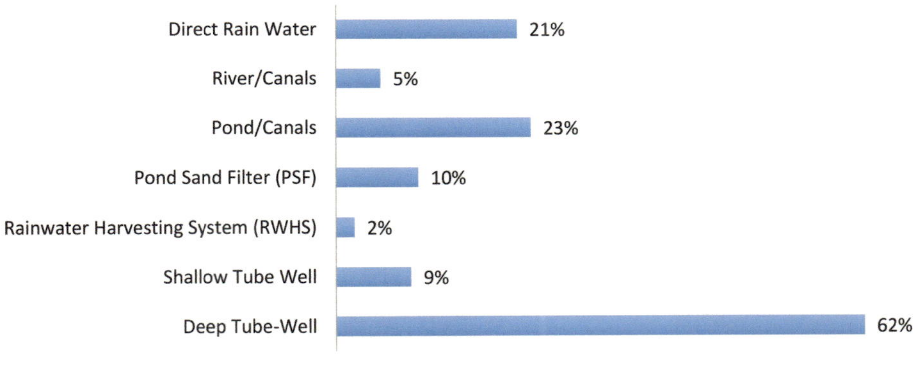

Figure 3.
Percentage of respondents showing their sources of drinking water supply in the study areas.

4.4 Salinity intrusion affects the sources of drinking water in the coast

It is important to note that salinity intrusion not only takes place at the groundwater level but also has been contaminating the surface water. Thus, **Figure 4** demonstrates the impacts of climate-induced disaster on the sources of drinking water. Wherever there is a cyclone like that of Aila that occurred in May 23, 2009 and Sidr in November 2007, a huge extent of the areas had been inundated extensively. The land was flooded, and saline water from the Bay of Bengal intruded the land and merged with the surface water. Nearly 35% of the respondents stated that due to the cyclone, the salinity of the water increases due to the intrusion of the bay water into the surface water, hence overlapping with drinking water. On the coastal areas, saline water also percolates down to the surface water and causes impacts on drinking capacity, and such statements have been stated by the majority of the respondent (as stated by 43% of the respondents). However, in such situations, ponds become a great resource for the coastal communities as a reservoir where freshwater could be collected and stored to be used for future instances. Waterlogging is a major issue when it rains heavily, and due to poor drainage system, the water gets collected on the land surface and that coincides with drinking water and causes climate hazard issues. This has been stated by 2% of the respondents in the study. Drought is a condition where the precipitation is below the average level which creates a crisis in drinking water. The sudden change of rainfall which is signified as erratic rainfall causes drinking water to be affected if it is massive or crisis if it is low. This has been stated by 29% of the respondents. The riverine flood causes intrusion of saline water into the river.

4.5 Impacts of climatic disasters on human life

The field study studied human perception to understand the impacts of disasters on people's lives. Through the survey analysis, it was found that 93% of the people acknowledged to agree that disasters make a huge negative difference to their livelihood, stability, and functioning, which eventually disrupts their sense of stability and rootedness to their origin. Disasters, especially which are of large scales, eventually compel the native communities to adapt to a new condition of living. Others choose to migrate to a potentially "safer" zone, where there could be better economic opportunities for living (**Figure 5**).

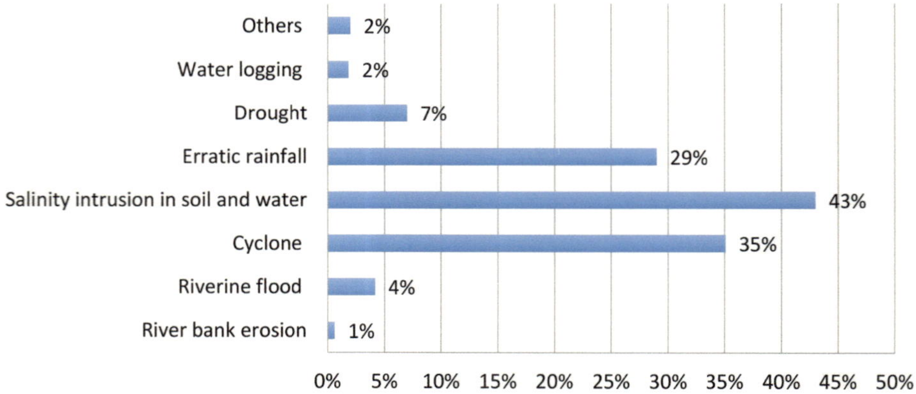

Figure 4.
Percentage of respondents of the impacts of climate induced disasters on the sources of drinking water.

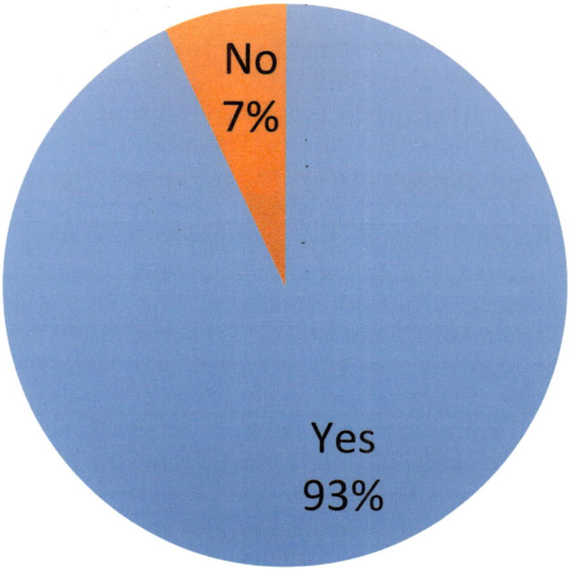

Figure 5.
Do climate disasters have any impact on your life?

4.6 Climate change, coastal livelihoods, and adaptation strategy

A number of climate-induced hazards have a colossal impact on the livelihood of the coastal people. Survey findings show that cyclone has the largest impact on the livelihoods of the communities, followed by salinity intrusion caused by storm surge. Excessive rainfall also has an impact on the livelihood of the coastal community people. It is observed that very small percentage of respondents in the survey mentioned about riverine flood, river bank erosion, drought, and sea-level rise as hazards that disrupt their livelihood.

The coastal area of Bangladesh has its unique characteristics. Communities living in the coastal zones have acquired their own techniques to combat the various climate adversities that hit their localities after a numbered return period or even erratically. However, there are some key roles and responsibilities of the government to ensure that the communities are far aware of the actions needed to be taken during and prior to the hitting of natural disasters. Coastal zone communities are particularly protected by the Sundarbans. However, the biodiversity is still at risk because of being an ecological hotspot within the country. It is imperative that the balance between environmental conservation and development is well understood by the community people to ensure that their adaptation approaches are sustainable in the long run. For this, the responsible authorities should emphasize on fostering knowledge on climate change and adaptation within the coastal zones in Bangladesh.

Coastal communities are at a threat to a number of natural disasters, making the roles of early warning extremely vital. The extent of effectiveness of the early warning systems should be revisited in the coastal zones so that there is sufficient lead time for people to prepare before flight and evacuation. The coastal zones of Bangladesh host a large number of people who contribute to the economic and agricultural sector of the country. These communities become stranded with the loss of lives of family members, livestock, and belonging. All of these could be avoided by a large scale if proper and timely early warnings are provided, which can be interpreted by the communities in the right way. Hence, training is essential for the people to communicate with the language of the early warning messages so that they can act in the right way. There should be adequate monitoring and evaluation on the condition of the shelters where people reside during the disaster period. This is important especially in the coastal areas of Bangladesh, where there is a need for sufficient lead time for people to move to a safer place prior to when disasters hit. On the other hand, livestock are important assets for the people who live in the coastal zones. The livestock should have veterinary services along with the doctors for the local people, both during and after disaster, as they are valuable assets for these marginalized people in the coastal areas.

The coastal zone communities have been largely benefitted from various trainings which has taken place in the past. This is particularly true for agriculture-based trainings. Audiovisuals can be used to capture the success stories and practices on agriculture, as well as challenges and failures, and should be channeled to the community through the local people. This increases the trust, motivation, and accountability among the community people to adopt the best practices and adapt with climate change.

Coastal livelihood in Bangladesh is largely dependent upon the access to utility services, livelihood, disaster preparedness and management, agriculture, inter alia, understanding the state of women's market access, and several other key factors. Women's involvement and access to the market are essential for the overall growth of the community to head toward resilience. In the coastal areas, salinity is one of the biggest concerns. To have proper access to market especially for women, the barriers to access to the market must be identified and addressed. After natural disasters take place in the coastal region, women have a big role to play to bring the family back to the state it was originally in. Identification of women and their barrier to access to the market may be done by the union with the help of the government or nongovernment organizations. Women's inclusion should be given greater regards including making key decisions and supply and demand of the essentials. Enhancing technical support and engaging local institutions into the grassroots-level advancements are crucial at the same time. In many communities, women are not allowed to make key decisions within the household. But women should be included in all forms of major decision-making processes which relate to disaster preparedness. Since the coastal zone is very susceptible to cyclones, the communities are constantly faced with instantaneous decision-making. The role of women is crucial in such decision-making processes. Hence, institutional approach to help include women in the key decision-making processes is essential, especially in a community like that of the coastal area of Bangladesh, where major decisions are taken by the male members of the family.

Coastal zone has been subjected to a number of big natural disasters in the past. The major cyclones that have hit the area are Cyclone Sidr in 2007 and Cyclone Aila in 2009. This has led to a large number of people to evacuate and move to a safer place. Such migration pattern has taken place in a number of stages, where the families have first moved further inland and then later were forced to move toward the city, where there are more income-generating opportunities. The Government of Bangladesh has been very particular to assessing the trends of migration patterns in Bangladesh within the last few years. As such, the patterns of migration should be traced from the place where they are displaced due to various extreme events or climate change impacts to the migrated area. Migration stories collected can be used for future purposes, and experiences gathered through these stories are assets to be used for future help of the people.

Local adaptation plans in the coastal zones should be strengthened by manifolds because it will help to address the problems of the local communities through locally

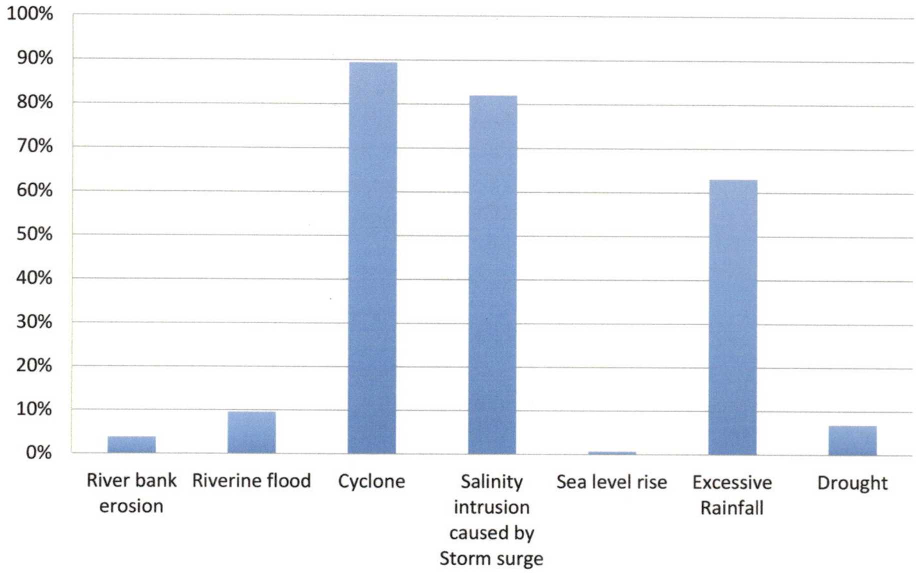

Figure 6.
Climate induced hazards that affect livelihoods of the study communities.

applied solutions. The coastal zone in Bangladesh has its own distinctive characteristic and properties. Area-specific problems must be identified for solutions, which too should be specific to each locality and their local problems. Within the coastal communities, the adaptation techniques should be in line with the region-based problem, which must be identified through further studies. This in turn will help not just the development sector but also the government to identify key strategies and solutions appropriate to lessen and manage local-level complications. Hence, various sectorial impacts should be assessed with scrutiny and be speculated with deeper interrogations and observation.

In the coastal zone of Bangladesh, gradual changes should be monitored and speculated with critical lens, of which cyclone is an example. Cyclone is a phenomenon which occurred quite a number of times in the coastal zone, and, hence, tracking its pathways and return periods is extremely important. Future studies should invest in monitoring how much has deviated in the timeliness of the events and its nature, i.e., gradual change between frequency, timeliness, and intensity of the extreme events, so that precautionary measures and subsequent activities can be executed duly with adequate knowledge. This will help to protect the mass number of people who are at risk of becoming homeless due to the onset of such large-scale disaster (**Figure 6**).

5. Conclusion

A compound impact of climate change and reduction of freshwater flow from the upstream is converting fresh water into saline zones. Climate change is enhancing the salinity problem in the coastal area of Bangladesh. Regional policies, agreements and strategies regarding trans-boundary issues need to be ratified.

The Bangladesh Water Development Board (BWDB) is responsible for maintaining and conducting the rehabilitation project of the polders. The BWDB has categorized 51 and 55 polders as most vulnerable and medium vulnerable, respectively. To address the vulnerability, it is important to repair the faulty infrastructure. Most of the sluice gates are damaged or broken which allow intrusion of saline water. Not only that, the shrimp farmers cut the embankment to get saline water into their shrimp fields which make the embankment weak. The weak embankment is easily damaged by tidal pressure, particularly during full moon, and thus saline water enters the polders. Regular maintenance of the embankments, dams, sluice gates, and polders are important to maintain the water balance. For this a robust monitoring system can be developed to ensure the accountability of the concern institutes and officials.

Bangladesh has already developed several salinity-tolerant rice varieties. However, with the increasing level of salinity investments, it should be directed toward research and development of saline water submergence-tolerant crop varieties and introduce the varieties in the rural areas. Lastly, livelihood options of the community people have to be ensured to develop their capacity to over any challenge.

Lastly, climate change must not be muddled with the innate geographical and hydrological disadvantage of Bangladesh that leads to cyclones and other disasters. Bangladesh hosts a combination of a myriad number of rivers, and the country also receives 1.2 billion tons of silt. There are other problems including arsenic in the water that many people drink which is not problem driven due to climate change. Climate change-related problems need to be specifically identified for each sector or thematic areas for each of the agroecological zones or hazard prone eco-zones for developing local adaptation solutions to local problems.

Author details

Golam Rabbani*, Sirazoom Munira and Samia Saif
Bangladesh Centre for Advanced Studies, Dhaka, Bangladesh

*Address all correspondence to: golam.rabbani@bcas.net

References

[1] Soil Resources Development Institute (SRDI). Saline Soils of Bangladesh. Dhaka, Bangladesh: SRDI, Ministry of Agriculture; 2010

[2] World Bank, 2000. Bangladesh: Climate Change & Sustainable Development. Report No. 21104 BD, Dhaka; 2000

[3] Global Economy. Available from: https://www.theglobaleconomy.com/Bangladesh/Share_of_agriculture/ Visited on 30 June; 2018

[4] Haque SA. Salinity problems and crop production in coastal regions of Bangladesh. Pakistan Journal of Botany. 2006;**38**(5):1359-1365

[5] Ahmad M, Jovel R, Narayan A, Mudahar MS, Ferdausi SA, Pusch C, et al. Damage, Loss, and Needs Assessment for Disaster Recovery and Reconstruction. Dhaka: International Development Community with Financial Support from the European; 2008

[6] Hofer T, Messerli B. Floods in Bangladesh. History, Dynamics and Rethinking the Role of the Himalayas. Tokyo: United Nations University Press; 2006. p. 468. ISBN: 92-808-1121-5

[7] Hossain MA. Global warming induced sea level rise on soil, land and crop production loss in Bangladesh. In: 19th World Congress of Soil Science, Soil Solutions for a Changing World, Brisbane; 2010. Available from: http://www.ars.usda. gov/Services/docs. htm

[8] Karim MF, Mimura N. Impacts of climate change and sea-level rise on cyclonic storm surge floods in Bangladesh. Global Environmental Change. 2008;**18**(3):490-500

[9] Habiba U, Abedin MA, Shaw R, Hassan AWR. Salinity-induced livelihood stress in coastal region of Bangladesh. In: Water Insecurity: A Social Dilemma. Bingley, United Kingdom: Emerald Group Publishing Limited; 2014. pp. 139-165

[10] Rawlani AK, Sovacool BK. Building responsiveness to climate change through community based adaptation in Bangladesh. Mitigation and Adaptation Strategies for Global Change. 2011;**16**(8):845-863

Chapter 7

Social Value of Urban Rooftop Farming: A Hong Kong Case Study

Ting Wang and Mathew Pryor

Abstract

As cities densify, areas available for agriculture within the city become increasingly small and infeasible for mass production. In parallel, many cities have seen a rapid rise in establishing community-based micro-farming, operating within marginal spaces of uncertain ownership or regulations. Prominently in Hong Kong, more than 60 urban rooftop farms have spontaneously appeared in the last 10 years on buildings. High application rates for renting plots in these informal farms suggest a strong demand in the population. Motivations cited by participants of rooftop farms are typically social, although social values have yet to be specifically defined or objectively measured. Hong Kong Special Administrative Region Government's new agricultural policy conceives urban agriculture as a commercially productive practice. In consequence, urban rooftop farming lies awkwardly between formal city planning and informal community practices. A study of five rooftop farms in Hong Kong found, through participant opinion surveys and cost-benefit analysis, that the social benefits to participants were multifaceted with a preference on personal socialization and that they were willing to pay for the experience. The results suggest that if the products of rooftop farming could be conceived as being social, rather than food production, individual motivations and state interests could be aligned and the available roof space activated to achieve a more sustainable city.

Keywords: urban agriculture, rooftop farming, social benefits, sustainability, cost-benefit analysis

1. Introduction

With the intense contest for ground-level space within high-density urban districts, urban agriculture has taken on multiple forms and occurs in different locations, such as peri-urban farming, urban soil-based farming, indoor farming and rooftop farming [1]. Urban agriculture was initially conceptualized as a response to increasing concerns for food security within the city, with the focus on the potential for mass production within a localized food system that includes production, processing, distribution, consumption and recycling [2]. More than 30% of the food requirements of the City of Oakland are planned to be provided from within the physical limits of the city through city council's sustainable food system [3]. However, within the complex morphology of high-density cities, the contest for space and strict land use and building controls, the large-scale contiguous spaces required for economic mass agricultural production are seldom available. Many micro-farming enterprises, however, have emerged in cities around the world as community gardens and allotment gardens [4]. Occupying small-scale,

IntechOpen

marginalized and fragmented "leftover" spaces, these occur on sites of uncertain ownership and ambiguous regulatory control.

A clear expression of this phenomenon is the spontaneous appearance in the last decade of more than 60 rooftop farms on underutilized flat roof spaces across the dense urban districts of Hong Kong [5]. These urban rooftop farms are composed of numerous lightweight surficial planter boxes (as opposed to the built-in planting constructions typical of green roofs) which are individually rented to the general public through community enterprise organizations or provided to relevant groups by corporate or institutional owners. Proximity to the people's living and working spaces have made urban rooftop farms popular, with all farms reporting that they are constantly heavily oversubscribed. Farm owners have suggested that the strong demand for participation is motivated by the opportunities it provides for social interaction, passive recreation, health, education and self-achievement. This contrasts with the HKSAR Government's recent policy initiatives for urban agriculture which are focused on economic and productive values [6]. In consequence, urban rooftop farms in Hong Kong are in an ambiguous situation between formal centralized city planning and informal community enterprise action. To understand the social benefits of rooftop farming within an urban context of contested space and extreme land value, this study looked to monetize social value through cost-benefit analysis and willingness among participants to pay for extra social benefits derived from the practice.

1.1 Social value of urban rooftop farming

Social value has long been a consideration within environmental justice discourses; however there has been relatively little research on the social values of urban agriculture and almost none on urban rooftop farming [7]. As with urban agriculture, the few policy debates that have occurred on urban rooftop farming have focused on the potential economic value—the monetary profits that might be generated by selling food produced within the city and generalized concerns for global food security. Around the world, however, very few large-scale commercial urban rooftop farms have been successfully established, and these have only been achieved by retrofitting rooftops with large-scale greenhouses, e.g., AeroFarms in the USA [8] and urban farmers in the Netherlands [9]. The large majority of urban rooftop farms have been small-scale social and community enterprises. In recent years, discussion about the practice has migrated onto to potential contribution to urban environment and greening [1, 10–13]. Urban rooftop farms have been suggested as possible patches that might visually and ecologically link existing green spaces and corridors within an integrated green infrastructure system and help mitigate urban heat island effects [14]. It has been shown that urban rooftop farms support far higher biodiversity (some have upwards of 200 plants species) than green roofs [15].

Only recently have discussions of the social values of urban rooftop farming begun to appear in the literature. Although social values are considered an important principle within broader concepts of urban sustainability, their recognition and development are lagging [16]. This is commonly attributed to the fact that social values associated with the external environment, such as green spaces and allotment gardens, are intangible and difficult to measure [17]. Social value is usually assumed to be generated through communal physical activity within a space, for example, social groups collaborating on planting activities [18]. Long return on investment makes social value hard to calculate and difficult to monetarize, metrics that are commonly required for inclusion in policy decision-making [19].

Through a review of international case studies, social values of urban rooftop farming were initially investigated from three aspects: social capital theory, landscape projects and urban agriculture practices and with the aim of building a systematic framework of social values for urban rooftop farming. As Dika and Singh [20] noted, the decomposition of a broad concept into factors and indicators can improve understanding and help the policy adaptation in specific contexts.

Ideas of social values are based on social capital theory which focuses on balancing different social groups by creating a sense of fairness from collaboration [21]. Social group integration and empowerment are key factors discussed by scholars. Dubos [22] suggests that social capital should be considered in two forms: structural network and cognitive value. Doherty further explains that the structural network in an inclusive society should cross generations and identities and consist of the behaviour-related indices of trust, informal networking, mutual support, reciprocity and solidarity [23, 24]. At the same time, cognitive value is a significant assessment for empowered citizens which is usually obtained from increasing self-satisfaction, achievement and leadership in the society [25, 26].

As an emergent landscape typology, performance measures for urban rooftop farming have yet to be developed [27]. Methods of measuring performance of built landscape have tended to assess physical objects and functional efficiency [28, 29]. Of the few approaches that have evaluated changes in social aspects, Landscape Performance Series (LPS) and Case Study Investigation (CSI) contain the most instructive framework, as they categorize recreation, health, education and food production as core social value factors that enhance sustainability in landscape projects [27].

In the absence of previous research on the specific social values of urban rooftop farms, this research drew upon discussions of social values related to urban agriculture in general. This allowed indicators for an urban rooftop farming social values framework to be identified. The urban agriculture matrices framework developed by Design Trust for Public Space program in New York highlighted the significant benefits through increased physical health and social empowerment from growing vegetables [30]. Specifically, physical, mental and dietary health can be summarized from the research outcome. Social empowerment has been further supported via environmental and food education, leadership and socializing activities which are increasingly important by-products of all forms of urban agriculture. Other researchers have identified unique collective social welfare being generated through urban rooftop farming [7, 31, 32]. Tian and Jim addressed the social value of additional open spaces to the surrounding communities through multifunctional roof spaces, noting that given the limited land in highly dense cities, retrofitting urban farms to rooftops can effectively activate large numbers of vacant spaces within the city for social benefits [32]. Prior research studies have also indicated that dynamic factors are involved in the generation of social values through the practice of urban agriculture.

1.2 Framework of social benefits of urban rooftop farming

Based on these interdisciplinary research studies, a social value framework for urban rooftop farming was developed, specific to the Hong Kong context (**Table 1**). This allows a spectrum of social benefits of urban rooftop farming to be considered, with respect to the diverse stakeholders' (state and individual) interests. The framework compares the social values generated by urban agriculture, green roof installations and rooftop farms; identified from published research papers; and categorized under six factors: health, education, community recreation, urban improvement, social empowerment and social group integration. Urban rooftop farming generates the greatest amount of activity across all the different social values.

URF social value framework					
Urban agriculture	**Green roof**	**Rooftop farms**	**Category**	**Factors**	**Social benefits**
√		√	Social benefits	Health	Improve physical health
√		√			Improve mental health
√		√			Experience health habit and diet
√		√		Education	Increase environmental awareness
√		√			Promote sustainable living
√		√			Increase organic food knowledge and demand
√		√			Gain practical skills by working in urban rooftop farms
√	√	√		Community recreation	Provide extra open space for communities
√	√	√			Provide visual aesthetic value
√	√	√			Increase space using comfortableness
√	√	√		Urban improvement	Serve as a planning tool to fill vacant spaces in cities
	√	√			Extension of the life expectancy of roofs
		√			Diverse the multifunctions of roof spaces
	√	√			Good for urban or building retrofitting
√	√	√		Social empowerment	Improve users/residents' life satisfaction
√		√			Enhance community participation
√		√			Develop leadership
√		√			Provide job opportunity to communities
√		√		Social group integration	Empower marginalized groups
√		√			Enrich aging life
√		√			Enhance parent and children relationship
√		√			Form social networks
√		√			Create social solidarity among diverse groups

Table 1.
Social value framework for urban rooftop farming.

2. Hong Kong urban rooftop farming in a high-density city

Within HKU's broad-based "edible roof" initiative which examined the rooftop farming phenomenon across Hong Kong, this specific research study examined eight urban rooftop farms within Hong Kong (including enterprise, social enterprise and individually oriented modes) to determine the nature and scale of the social values that urban rooftop farms could generate.

Hong Kong is an extreme example of high-rise high-density urban settlement, with severe contest for ground-level space, very high land values and a passive governance structure. Although HKSAR Government's New Agricultural Policy 2014 and Hong Kong 2030+ strategic planning statement do acknowledge urban rooftop farming practices within the general concept of urban agriculture, intention has focused primarily on economic productivity, and no specific institutional, regulatory or technical support is offered to the small-scale grassroot organizations that practice farming. Despite this, more than 60 urban rooftop farms have spontaneously appeared in the city since 2008 covering some 15,000 sqm of previously underutilized roof space [5]. The majority of the farms are located on industrial or institutional buildings within the older urban districts (**Figure 1**). Based on a definition of the physical and operational limits of rooftop farming practices and subsequent suitability assessment of all existing buildings in the territory, the potential farmable roof spaces that might exist within the city have been estimated at approx. 595 ha [5]. Although typically small-scale and disparate, these spaces are all in close proximity to large urban populations and collectively offer an expansive opportunity for generating social value (and its attendant economic advantages) if activated for rooftop farming [33].

Physical and operational characteristics of the three modes of urban rooftop farming in Hong Kong were identified through systematic site survey and typological study (**Figure 2**). Social enterprise farms aim to promote social change through a sustained commercial business [34]. Social enterprises, such as City Farm and Fun n Farm, generate social impacts by renting out the planting plots to the public. Planting plots typically consist of shallow free-standing black plastic crates filled with lightweight soil, with bamboo or plastic pipe frames above supporting screen netting [33, 35]. Crops are selected and taken care of by farmers, although daily watering is undertaken by farm managers. Training courses (for different skill levels) and related social and craft activities are commonly offered. Farmers rent any number of boxes per month, depending on their ambition and commitment. All farms report extensive

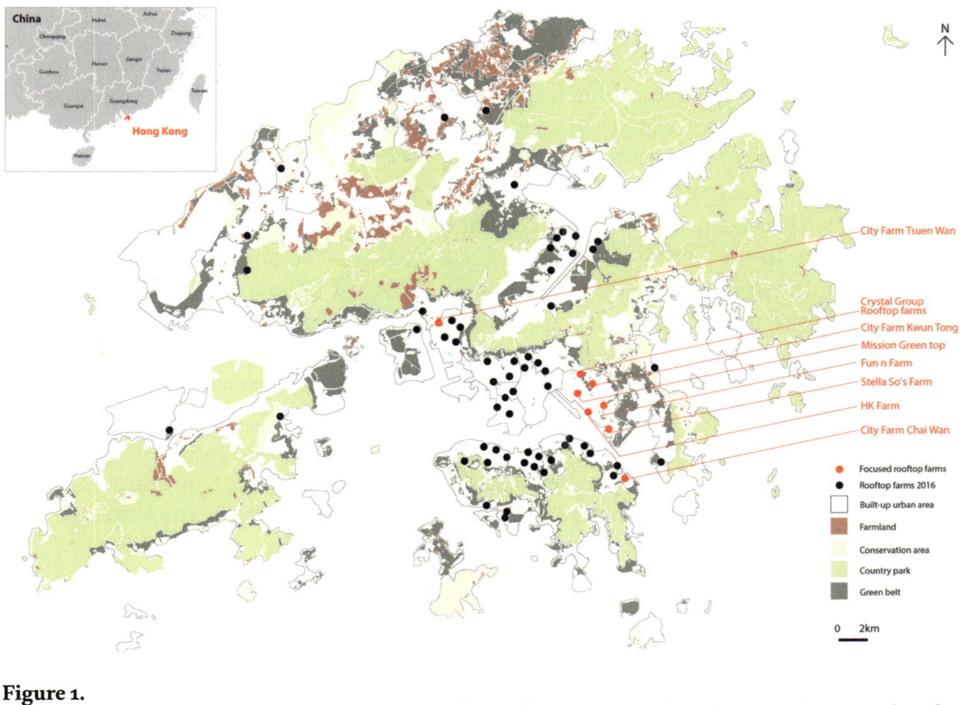

Figure 1.
Locations of urban rooftop farms in Hong Kong, as of 2016 (data source: Mathew Pryor ongoing research and Google earth).

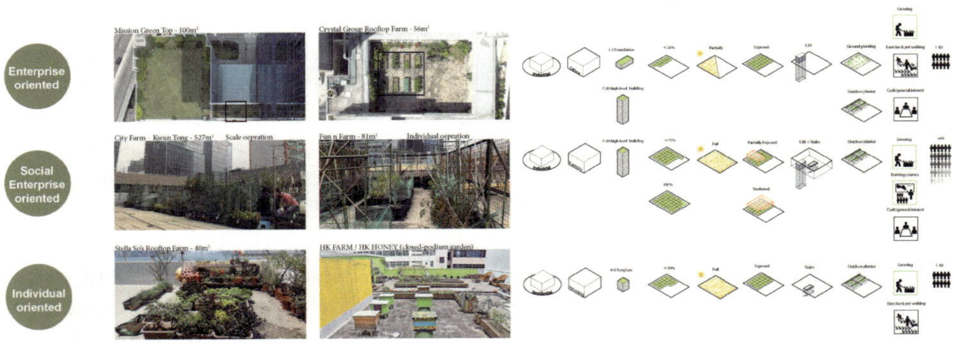

Figure 2.
Typological study of urban rooftop farm in Hong Kong (photo taken by Mathew Pryor and ting Wang).

waiting lists. The depth of soil and exposure to wind limits species choice to some extent, but a wide range of leafy greens, climbing plants, root vegetables and herbs can be grown. Enterprise-oriented farms are operated by private companies and business or large institutions (universities, schools, hospitals) located on their own premises. Access to the farms is restricted to employees or institutional members. They are similar in physical form and nature to social enterprise farms but additionally provide leisure and social space for employees, with tables, chairs, etc. Individual rooftop farms were very small-scale and only found on residential buildings. Their form was typically more complex and less ordered, and both the form of the planter and the crop species were far more diverse. As they depend solely on the individual owner's willingness and availability, they were seen as being more vulnerable.

3. Research design

Based on this understanding of the local context, the research study was structured around a participant opinion survey and semi-structured interviews with the operators from five selected farms. The survey aim was to validate the preliminary urban rooftop farming social value framework and to quantify the intangible social values from the perspective of the users, including those with and without experience of farming. Subjects were randomly selected from the five farms and from the surrounding residential communities, respectively. A total of 108 answers were collected.

Semi-structured interviews were conducted with farm operators from the five farms, in order to understand the monetary influence of social values in urban rooftop farming and to verify the findings from survey. Questions focused on topics such as modes of operation and costs, as well as physical arrangement and planting types. Farm cost data was used in cost-benefit analysis and "willingness to pay" based on contingent valuation methods and perception preference methods. As willingness to pay is influenced by the perceived utility, personal preference of use and socioeconomic environment of the subjects, the survey was designed to obtain the information about various degree of willingness and payments, preference of social values developed in framework and personal socioeconomic information including gender, employment, education and income levels.

4. Findings

The majority of respondents (77%) perceived social values to be the most important benefits of urban rooftop farming, compared with environmental values

(58%) and economic benefits (10%). Women and the middle-aged (30–50) were found to be the predominant users of urban rooftop farms—by both number and time. This finding was confirmed through farm membership records and observations of farm managers. Meanwhile, the majority of farm participants were from middle- to high-income groups.

4.1 Social values with a preference for personal socialization

The perception of social values was complex, with individuals expressing degrees of perception toward the six different factors (**Figure 3**). However, personal socialization benefits were identified most strongly among the six factors. Health (53%) and education (62%) were the factors most perceived by respondents that directly link to the personal enhancement in social statues. Planning social welfare (40%), social group integration (40%), community recreation (35%) and social empowerment (25%) were of less importance by respondents.

Disparity of social benefit preferences reflects the difference between personal experience values and group conceptual values. Personal health and education are the most direct feelings obtained through daily activities; however, individuals perceive larger scale community and collective benefits indirectly. For instance, though social group integration was not perceived as very significant on the whole, the indicators for enriching the life of the aged and enhancing intergenerational relationships were perceived as highly significant because of the close personal feelings attached. "Developing leadership" and "providing job opportunities"

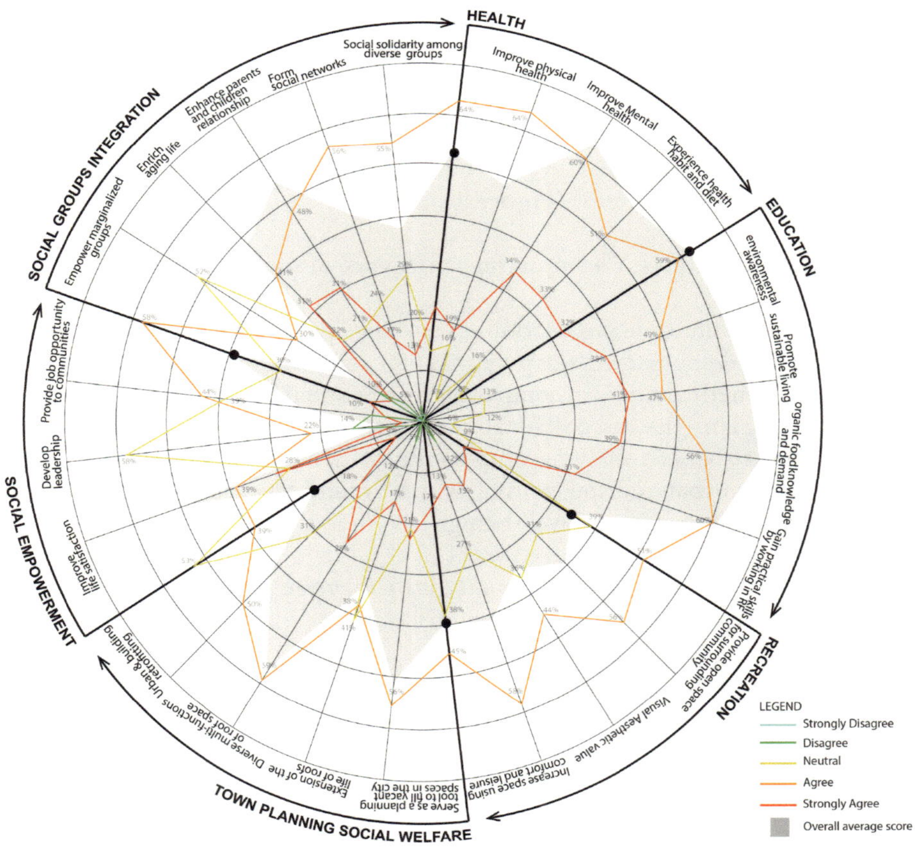

Figure 3.
Perception distribution.

were the two least important indicators among the social empowerment factors, in interview participants questioned "how can leadership be improved by just growing vegetables?" To some extent, this makes sense because it is hard for leadership development to be perceived by the users themselves unless there is an external instructor who guides the activity and highlights the purposes behind it. This may necessitate long-term observation of farm participation organized by experienced teams or working feedback from the employment company. At this point, there is no measurable index for conceptual benefits. In addition, the benefits of increased job opportunities for the society will only be realized when urban rooftop farming becomes a city-scale endeavor. Current rooftop farms are individually too small to be measured in the employment indicator.

4.2 Willingness to pay for social benefits

Many respondents indicated willingness to pay for the social value experience derived from urban rooftop farming. While some were conservative about payment, "I don't have extra time to enjoy the rooftop farms" (32%); "I cannot afford to pay or buy the service" (19%), the majority of respondents (87%) were willing to pay. The average payment reported during the survey was HK$ 220 per month/person/half square meter. In comparison with the current charge for renting a plot in an urban rooftop farm (HK$ 190), this suggested an increased perception of social values among users.

Just asking questions about individual payment decisions encouraged respondents to consider the benefits and the maximization of utility. Willingness to pay was found to be related significantly to the degree of understanding of urban rooftop farming, level of education and income level. Willingness to pay increased with the cognitive level of participants from "no idea" to "have participated in urban rooftop farming." Practicing farmers were willing to pay more (HK$ 232) than those that had not previously participated (HK$ 194). Most of the respondents who are willing to pay were from higher levels of education (undergraduates and graduates), as well as higher-income groups (**Figure 4**).

4.3 Cost-benefit analysis: The monetary influence of social values

Apart from the multiple implications of social values in urban rooftop farming, this research also demonstrates the potential monetary influence through the application of cost-benefit analysis in comparing the marginal benefits (social values) with the existing benefits and costs (capital and recurrent). According to [25, 36], the following cost-benefit analysis components can provide an economic spectrum of social values in urban rooftop farming which can influence government decision-making and contribute to social well-being:

$$\text{Gross costs} = \textit{Capital costs} + \textit{Recurrent costs} \tag{1}$$

$$\text{Gross benefits} = \textit{Recurrent benefits}\ (\textit{or any other marginal benefits}) \tag{2}$$

$$\text{Net Benefit} = \textit{Gross benefits} - \textit{Recurrent costs} \tag{3}$$

$$\text{Cost and Benefit ratio} = \frac{\textit{Gross Benefits}}{\textit{Gross Costs}} \tag{4}$$

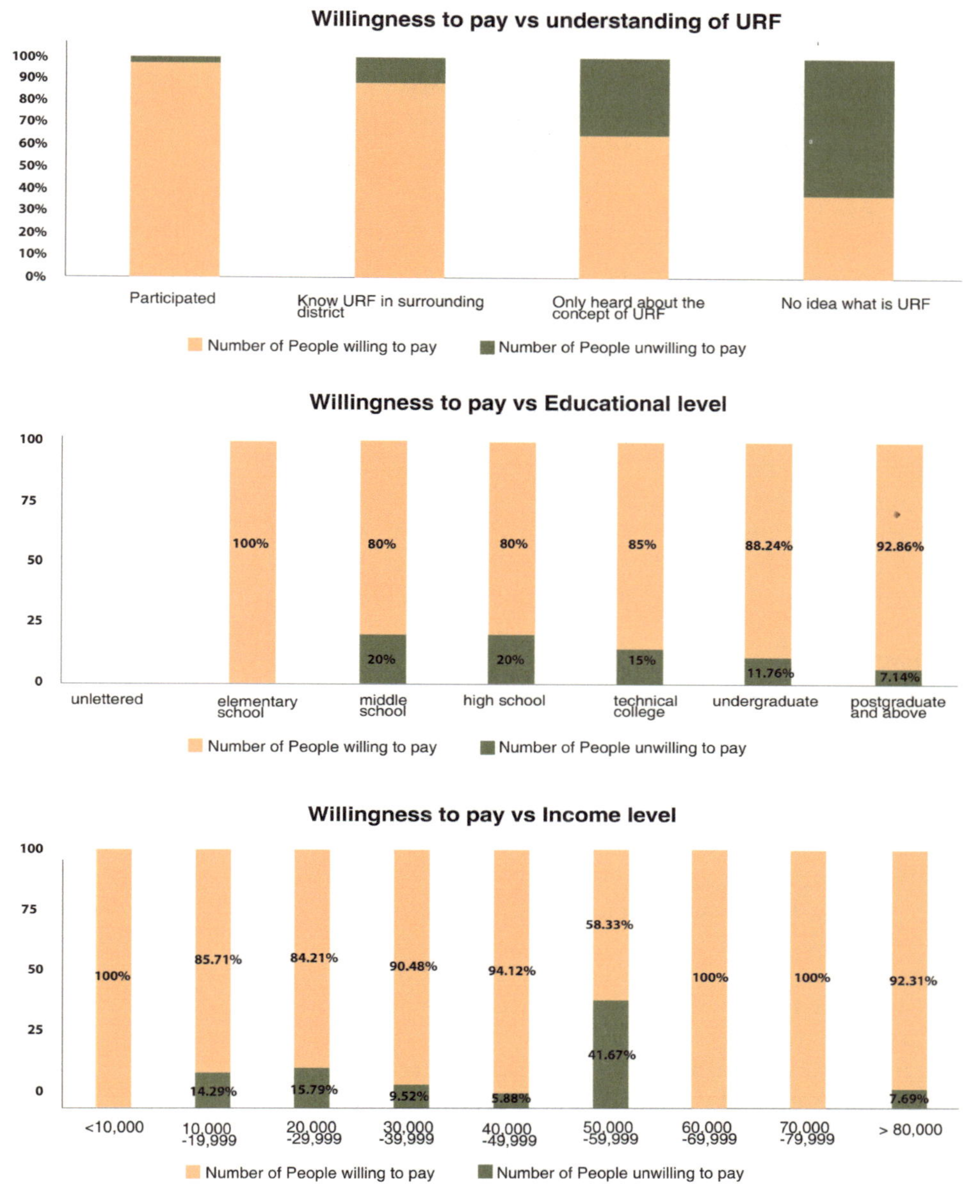

Figure 4.
Significant factors in willingness to pay.

$$\text{Payback Time} = \left(\frac{1}{\textit{Cost and Benefit ratio}}\right) * \textit{one year} \tag{5}$$

$$\text{Net Present value} = \frac{\left(\textit{Net Benefit} * (1 - (1 + \textit{Discount Rate}))^{-\text{project period}}\right)}{\textit{Discount Rate}}\ \textit{Capital costs} \tag{6}$$

Among the financial information obtained from operators, City Farm Kwun Tong was chosen as a prototype for this calculation due to its comprehensive operational mode and representativeness of other farms in Hong Kong. Cost-benefit analysis in the study used the basic scenario of a rooftop farm in Hong Kong. The prices and amounts were all generic estimates in order to provide the minimum costs and benefits.

First year revenues generated through urban rooftop farming were found to barely offset the costs in Hong Kong. In the prototype case, the gross costs and benefits of urban rooftop farming in the first year were HK$ 730,400 and HK$ 764,760, respectively. In subsequent years, the annual recurrent commercial benefits exceeded the annual recurrent costs HK$ 530,400, giving a benefit-cost ratio of 0.32 (234,360/7,304,000), which suggested a likely payback period of 38 months. This factors in the high initial capital cost to establish a rooftop farm which includes building retrofitting costs and the purchase of equipment. Farm managers reported that the business stabilized after the second year and revenues were expected to increase in a long term.

However, the current amount of payment is based on a narrow view of farming participation (HK$ 190 per month/person/half square meter). As suggested by the willingness to pay analysis, once participants took into account the social values derived from their farming activities, they might be willing to pay more (HK$ 220). If fees were raised to this level, it would significantly alter and increase the gross benefits (to HK$ 872,760 per year) and shorthorn the payback period (to 26 months). The results suggest that cost-benefit analysis provide a useful basis on which to reconceive the financial viability of the urban rooftop farms.

5. Discussion

In Hong Kong, formal green initiatives in the urban area have come a long way from the development of public parks in the 1970s to the promotion of green roof designs through sustainable building directives in the 2000s. However urban rooftop farming has not been formally recognized and exists still within gray areas of urban planning legislation and building control.

As evidenced by these findings, the disparity of multifaceted social values aligns with previous literature on social capital theory. Cognitive values are directly related to the individuals in the society such as the effects of health and education improvement, while structure values are indirectly built through expanding network in society which needs more efforts to achieve. For instance, collective assets like the urban economy prosperity and social solidarity not only improved by mobilizing individuals through urban rooftop farming but also need more complex catalysts.

Different levels of understanding of social values have been identified within previous landscapes value research [37]. Individual perceived values in the landscape, concentrating on health and general wellbeing, have most readily been identified: collective values relating to spatial planning and resource management have been less mentioned by subjects. This disparity is also rooted in the physical nature of existing urban rooftop farming practices. According to observations made during this research, rooftop farming activity is explicitly individual due to space limitations. A large number of planting plots were arranged side by side within physically constrained roof spaces, inhibiting interaction. Participants work by themselves on individual plots while only "keeping an eye" on surrounding plots farmed by others. This mode of operation might explain the higher perception of direct personal health and education benefits. The lack of additional social space in social enterprise farms and the solitary nature of individual farms may reduce perceptions of collective social value such as engagement of the community or improvement of the urban environment.

Previous research has not explored the monetary influence of social benefits, which is required for urban rooftop farming to be incorporated into urban policy-making. For instance, on average the payback period for farms is shorter than for green roofs in Hong Kong (27 months) and for ground-level urban agriculture projects (96 months) [25, 38]. The monetary influence of social values is likely to become amplified as urban density increased. Governments, as well as building owners, are likely to be more willing to invest in urban rooftop farming for both the economic benefit and social value through community sustainability.

6. Conclusion and further research

A shift in the thinking about the products of urban rooftop farming from food security and urban greening to social benefits and positive support to activate urban rooftop spaces would create significant opportunities for aligning individual motivations and state interests, thereby achieving a more sustainable city. Though current urban rooftop farming is undertaken by individuals and grassroots organizations, with limited policy or technical supports from city authorities, users still perceived considerable social benefits in the form of sustainable living, environmental knowledge and enhanced relationships within social groups. Users' willingness to pay for the experience indicates that urban rooftop farming is a passive social activity which can be enhanced by collaborative activities and by-products of farming which include talking, working side by side, standing and comparing.

The implication of the multifaceted social values of rooftop farming suggests a changing perception of urban agriculture. With the increasing speed of urban densification, urban agriculture, constituted by complex social values and diverse interests from stakeholders, has the capacity to be a public good for cultural exchange and enhancing social coherence. This changing perception suggests the need for greater stakeholder support, recognition in legislation and integration with urban planning and building control processes. As an emerging urban activity, further studies are required. For instance, the higher preference for health and education as social benefits in this research requires more specific study to develop detailed instruments for those single indicators within particular groups. In addition, as this study only addressed the social values of urban rooftop farming in Hong Kong, further studies in different contexts and forms could help to expand the urban agriculture discourse.

Acknowledgements

The authors would like to thank the staffs from City Farm, Rooftop Republic Urban Farming as well as an anonymous friend for their generous assistance in the data collection of this research. This research received publication fund from the University of Hong Kong Department of Architecture.

Appendices

List of semi-structured interview questions:

1.1. When was your rooftop farm built?

1.2. How was your urban rooftop farm established? What kind of costs is included in the farm? Can you give me the rough number about the cost?

1.3. How does your farm operate on a daily basis? How many people did you hired and in what position? What kind of benefits can be earned in the urban rooftop farm? Can you give me the rough number about the benefits?

1.4. What are the difficulties you faced when setting up an urban rooftop farm in Hong Kong?

2.1. How big is your urban rooftop farm?

2.2. What kind of activities you have in your farm?

2.3. Can you estimate roughly how many people come to your rooftop farm on a regular basis?

3.1. What kind of species can you grow in your rooftop farm?

4.1. How do you think about the distribution characteristics of the participants in my questionnaire? Is it consistent with your observation every day?

4.2. How do you think about the existing result of questionnaire that shows the low perception of the collective social value in URF? Are you considering to add more public spaces or people to socialize in the future?

Author details

Ting Wang* and Mathew Pryor
The University of Hong Kong, Hong Kong

*Address all correspondence to: sarahwin@connect.hku.hk

References

[1] Sanyé-Mengual E et al. Resolving differing stakeholder perceptions of urban rooftop farming in Mediterranean cities: Promoting food production as a driver for innovative forms of urban agriculture. Agriculture and Human Values. 2016;**33**(1):101-120

[2] Smit J, Nasr J, Ratta A. Urban Agriculture: Food, Jobs and Sustainable Cities. New York, USA: The Urban Agriculture Network, Inc.; 1996. pp. 35-37

[3] Green M. Oakland Looks toward Greener Pastures T.O.F.P. Council. Oakland: Edible East Bay; 2007. pp. 36-37

[4] Van Veenhuizen R. Cities farming for the future. Cities farming for future. In: Urban Agriculture for Green and Productive Cities. The Netherlands: RUAF Foundation, IDRC and IIRP, ETC-Urban agriculture, Leusden; 2006. pp. 2-17

[5] Hui S. Green roof urban farming for buildings in high-density urban cities. In: The 2011 Hainan China World Green Roof Conference. Hainan,China: The University of Hong Kong; 2011

[6] Bureau F.a.H., F.a.C.D. Agriculture. The New Agricultural Policy: Sustainable Agricultural Development in Hong Kong. 2014 [cited 2018]; Available from: https://www.gov.hk/en/residents/government/publication/consultation/docs/2015/Agricultural_Policy.pdf

[7] Peck SW et al. Greenbacks from Green Roofs: Forging a New Industry in Canada. Toronto: Citeseer; 1999

[8] AeroFarms. Our Story. 2019 [cited 2018]; Available from: https://aerofarms.com/story/

[9] Space&matter. Urban Farmers in the Hague. 2016 [cited 2017]; Available from: http://www.spaceandmatter.nl/urbanfarmers

[10] Howarth RB, Farber S. Accounting for the value of ecosystem services. Ecological Economics. 2002;**41**(3):421-429

[11] Lin BB, Philpott SM, Jha S. The future of urban agriculture and biodiversity-ecosystem services: Challenges and next steps. Basic and Applied Ecology. 2015;**16**(3):189-201

[12] Cohen N, Wijsman K. Urban Agriculture as Green Infrastructure: The Case of New York City. Vol. 27. The Netherlands: Urban Agric; 2012. p. 16

[13] Peng LL, Jim CY. Economic evaluation of green-roof environmental benefits in the context of climate change: The case of Hong Kong. Urban Forestry & Urban Greening. 2015;**14**(3):554-561

[14] Borysiak J, Mizgajski A, Speak A. Floral biodiversity of allotment gardens and its contribution to urban green infrastructure. Urban Ecosystem. 2017;**20**(2):323-335

[15] Lélé SM. Sustainable development: A critical review. World Development. 1991;**19**(6):607-621

[16] Zahina-Ramos JG. Attitudes and perspectives about backyard food gardening: A case study in South Florida. In: The Charles E. Schmidt College of Science. USA: Florida Atlantic University; 2013. p. 125

[17] Van Leeuwen E, Nijkamp P, de Noronha Vaz T. The multifunctional use of urban greenspace. International Journal of Agricultural Sustainability. 2010;**8**(1-2):20-25

[18] Ganguly S et al. Lively'Hood Farm Financial Analysis a Feasibility Study

of Commercial Urban Agriculture in the City of San Francisco. Created by: Team Lively'Hood. San Francisco: SF Environment; 2011

[19] Dika SL, Singh K. Applications of social capital in educational literature: A critical synthesis. Review of Educational Research. 2002;**72**(1):31-60

[20] Forrest R, Kearns A. Social cohesion, social capital and the neighbourhood. Urban Studies. 2001;**38**(12):2125-2143

[21] Dubos R. Social Capital: Theory and Research. New York: Routledge; 2017

[22] Doherty K. Urban Agriculture and Ecosystem Services: A Typology and Toolkit for Planners. Amherst: University of Massachusetts Amherst; 2015

[23] Krishna A, Uphoff NT. Mapping and Measuring Social Capital: A Conceptual and Empirical Study of Collective Action for Conserving and Developing Watersheds in Rajasthan. India: World Bank, Social Development Family, Environmentally and Socially Sustainable Development Network; 1999

[24] Fan Y-w. Cost-benefit analysis of green roof application in telecommunication building in Hong Kong. [HKU theses Online (HKUTO)]; 2016

[25] Chui L. Skyrise greenery development in the Hong Kong context. [HKU theses Online (HKUTO)]; 2015

[26] Luo Y, Li M-H. A study of landscape performance: Do social, economic and environmental benefits always complement each other? Landscape Architecture Frontiers. 2014;**2**(1):42-57

[27] Rogler K. Data Farming: Demonstrating the Benefits of Urban Agriculture. 2013 [cited 2018]; Available from: http://thisbigcity.net/data-farming-demonstrating-the-benefits-of-urban-agriculture/

[28] Rahman SRA et al. Perception of green roof as a tool for urban regeneration in a commercial environment: The secret garden, Malaysia. Procedia-Social and Behavioral Sciences. 2015;**170**:128-136

[29] Tian Y, Jim C. Development potential of sky gardens in the compact city of Hong Kong. Urban Forestry & Urban Greening. 2012;**11**(3):223-233

[30] Pryor M. The Edible Roof: A Guide to Productive Rooftop Gardening. Hong Kong: MCCM Creations; 2016

[31] Borzaga C, Defourny J. The Emergence of Social Enterprise. Vol. 4. London: Psychology Press; 2004

[32] Hui M. In Organic-Hungry Hong Kong, Corn as High as an Elevator's Climb. The New York Times: New york; 2012

[33] Kahneman D, Knetsch JL. Valuing public goods: The purchase of moral satisfaction. Journal of Environmental Economics and Management. 1992;**22**(1):57-70

[34] Allen IE, Seaman CA. Likert scales and data analyses. Quality Progress. 2007;**40**(7):64

[35] Government HK. Kwun Tong District Profiles. 2017 [cited 2018 03/04]; Available from: https://www.bycensus2016.gov.hk/en/bc-dp.html

[36] Ahmed US, Gotoh K. Cost-Benefit Analysis of Environmental Goods by Applying Contingent Valuation Method. Tokyo: Springer; 2006

[37] Fagerholm N, Käyhkö N. Participatory mapping and geographical patterns of the social landscape values of rural communities in Zanzibar,

Tanzania. Fennia-International Journal of Geography. 2009;**187**(1):43-60

[38] Kim JS et al. Food and the City: A Smarter and Greener Approach towards Urban Regeneration - A Case of Suzhou Industrial Park, China. Shanghai: Xuelin Publishing House; 2018